EXPERIMENTS FOR INTRODUCTION TO
ORGANIC CHEMISTRY

A MINISCALE APPROACH

Frederick A. Bettelheim

Joseph A. Landesberg

Adelphi University

BROOKS/COLE

CENGAGE Learning

Australia • Brazil • Japan • Korea • Mexico • Singapore • Spain • United Kingdom • United States

BROOKS/COLE
CENGAGE Learning

For product information and technology assistance, contact us at
Cengage Learning Customer & Sales Support, 1-800-354-9706.

For permission to use material from this text or product, submit all requests online at **www.cengage.com/permissions**. Further permissions questions can be emailed to **permissionrequest@cengage.com**.

ISBN-13: 978-0-03-019238-8
ISBN-10: 0-03-019238-2

Brooks/Cole
10 Davis Drive
Belmont, CA 94002-3098
USA

Cengage Learning is a leading provider of customized learning solutions with office locations around the globe, including Singapore, the United Kingdom, Australia, Mexico, Brazil, and Japan. Locate your local office at: **www.cengage.com/global**.

Cengage Learning products are represented in Canada by Nelson Education, Ltd.

To learn more about Brooks/Cole, visit **www.cengage.com/brookscole**.

Purchase any of our products at your local college store or at our preferred online store **www.ichapters.com**.

Printed in the United States of America
7 8 9 10 11 12 13 12 11 10 09

Preface

This laboratory manual was written for an introductory, usually one-semester or two-quarter course, in organic chemistry. A number of the experiments are adapted from our highly successful text, *Laboratory Experiments for General, Organic, and Biochemistry,* second edition. Each of the experiments was designed with the following ideas in mind: (a) the experiments illustrate the concepts learned in the lecture part of the course; (b) they are clearly and concisely written so that students can easily understand them; (c) the experiments can be performed in a 2 to 2 1/2 hour lab period; (d) the chemicals used in the experiments are on a mini scale; (e) hazardous chemicals are kept to a minimum; and (f) the costs of running the experiments are kept low.

We chose the mini scale in order to provide economy as well as to reduce the disposal of the products of the experiments. In addition, the mini scale reduces students' exposure to hazardous chemicals, while still providing reproducible results.

The 32 experiments in this book provide a suitable choice for the instructor who usually selects 12-14 experiments for the one-semester course. Of these experiments, six illustrate experimental techniques, two deal with the manipulation of molecular models, eight describe the identification and properties of organic compounds, six pertain to the isolation and identification of compounds from natural sources, and ten provide synthesis. These experiments all have been thoroughly class tested at Adelphi University.

ORGANIZATION

Experiments 1 and 10 deal with the study of structure by allowing students to manipulate molecular models. Experiments 2-5 illustrate general laboratory techniques that students will use throughout the rest of the course. In Experiments 5 and 6 students learn isolation techniques commonly used in obtaining a compound from natural sources. Experiments 7-17 investigate the properties of different groups of compounds, progressing from hydrocarbons, to alcohols, to aldehydes, to carboxylic acids, and to amines. In a few experiments (8 and 11) the mechanism of the particular reaction is also explored. Experiments 12, 13, 18, 19, 22, and 23 demonstrate different methods of organic synthesis, progressing from the simple (Experiment 12) to the multistep, more complex route (Experiment 22). In the second half of the book many experiments pertain to reactions that are important from a biochemical point of view. Carbohydrates (Experiment 16), lipids (Experiments 26-28), proteins (Experiments 30-31), and nucleic acids (Experiment 32) are the main areas of this introduction, but one experiment (24) is designed to acquaint students with enzyme-catalyzed reactions.

FEATURES

- Each experiment begins with background information in which all the relevant chemical principles and applications are reviewed. This is not a repetition of textbook material but an adaptation to the concepts at hand.
- The procedures are step-by-step descriptions of the experiments. Special emphasis was made to make this section concise and clear for the reader.
- Caution! signs alert students to proceed with care. Some chemicals are poisonous at high concentrations. When using these substances, the amounts and time exposure are minimized to avoid health hazards.
- Pre-laboratory questions are provided with each experiment to familiarize students with the concepts and procedures before they start the experiment. By requiring students to answer these questions and by grading their answers, instructors can make sure students are better prepared for the experiments.
- The data report sheets ask students to record the raw data of the observations and to interpret the results of the experiments.
- The post-lab questions are designed for students to reflect upon the results and their significance.
- A brief introduction to infrared and nuclear magnetic resonance spectroscopy is found in the appendices. Also in the appendices are detailed instructions on how to prepare the solutions and other chemicals and how much material and what equipment is needed for a class of 25 students.

ACKNOWLEDGEMENTS

These experiments have been used by us and our colleagues over many years. We thank Stephen Goldberg, Jerry March, Sung Moon, Reuben Rudman, Charles Shopsis, and Stanley Windwer for their advice and helpful comments. Their expertise was instrumental in refinement of the experiments.

We also thank the following individuals who reviewed our manuscript: Kenneth K. Andersen, University of New Hampshire; Ardeshir Azadnis, Michigan State University; Dana S. Chatellier, University of Delaware; Sally Jacobs, University of Maine; and Richard T. Luibrand, California State University, Hayward.

Finally we extend our appreciation to the entire staff at Saunders College Publishing, especially to John Vondeling, our Publisher, and Sandi Kiselica, our Development Editor, for their encouragement and excellent efforts in producing this book.

We hope that you find our book of laboratory experiments helpful in the instruction of students and that the students enjoy performing the experiments.

Frederick A. Bettelheim
Joseph Landesberg
August 1996

*This book is dedicated to our wives: Vera S. Bettelheim and Lucy Landesberg
whose help, understanding and patience enabled us to write this book.*

Table of Contents

Experiment

Appendix

GLASSWARE AND EQUIPMENT COMMONLY USED IN THE ORGANIC LABORATORY

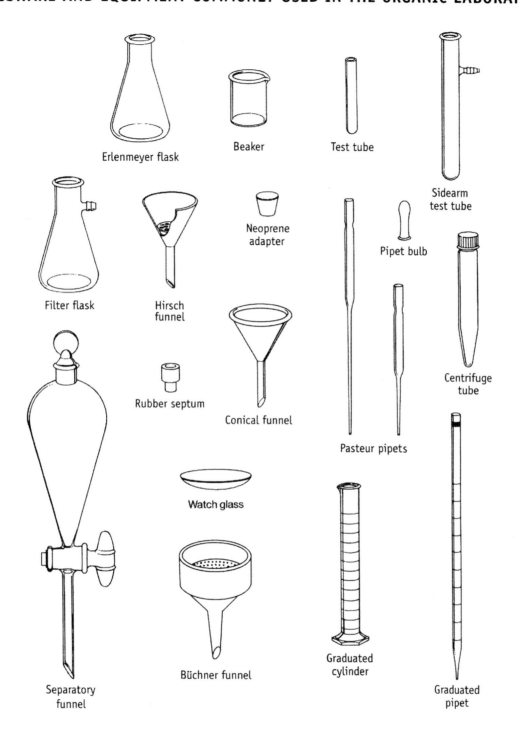

Erlenmeyer flask

Beaker

Test tube

Sidearm
test tube

Filter flask

Hirsch
funnel

Neoprene
adapter

Pipet bulb

Rubber septum

Conical funnel

Pasteur pipets

Centrifuge
tube

Watch glass

Separatory
funnel

Büchner funnel

Graduated
cylinder

Graduated
pipet

GLASSWARE AND EQUIPMENT COMMONLY USED IN THE ORGANIC LABORATORY

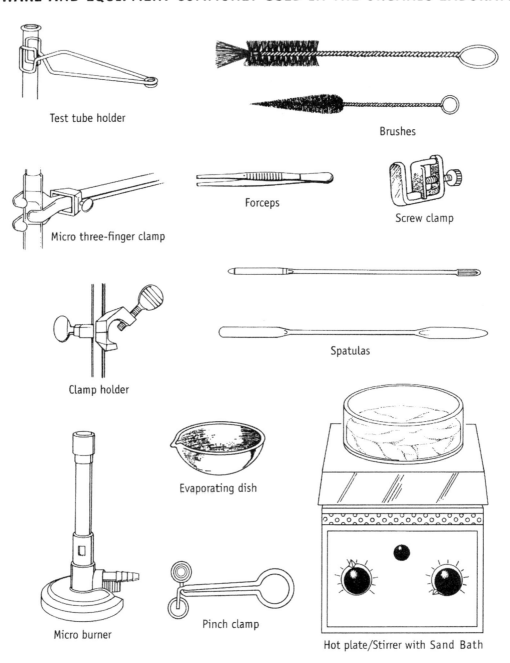

Test tube holder

Brushes

Micro three-finger clamp

Forceps

Screw clamp

Clamp holder

Spatulas

Micro burner

Evaporating dish

Pinch clamp

Hot plate/Stirrer with Sand Bath

GLASSWARE AND EQUIPMENT COMMONLY USED IN THE ORGANIC LABORATORY

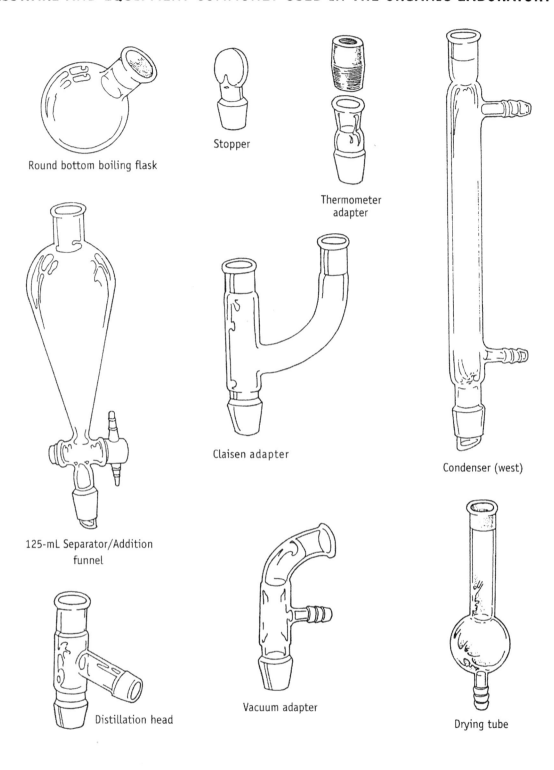

Round bottom boiling flask

Stopper

Thermometer adapter

Condenser (west)

125-mL Separator/Addition funnel

Claisen adapter

Distillation head

Vacuum adapter

Drying tube

x

EXPERIMENT

1

Structure in Organic Compounds: Use of Molecular Models. I

BACKGROUND

The study of organic chemistry usually involves those molecules which contain carbon. Thus, a convenient definition of organic chemistry is the chemistry of carbon compounds.

There are several characteristics of organic compounds that make their study interesting:

a. Carbon forms strong bonds to itself as well as to other elements; the most common elements found in organic compounds, besides carbon, are hydrogen, oxygen and nitrogen.

b. Carbon atoms are generally tetravalent. This means that carbon atoms in most organic compounds are bound by four covalent bonds to adjacent atoms.

c. Organic molecules are three dimensional and occupy space. The covalent bonds which carbon makes to adjacent atoms are at discrete angles to each other. Depending on the type of organic compound, the angle may be 180°, 120°, or 109.5°. These angles correspond to compounds which have triple bonds (1), double bonds (2) and single bonds (3), respectively.

$$-C\equiv C- \qquad\qquad \text{\textbackslash}C=C\text{/} \qquad\qquad -\overset{|}{\underset{|}{C}}-\overset{|}{\underset{|}{C}}-$$

(1)   (2)   (3)

d. Organic compounds can have a limitless variety in composition, shape and structure.

Thus, while a molecular formula tells the number and type of atoms present in a compound, it tells nothing about the structure. The structural formula is a two-dimensional representation of a molecule and shows the sequence in which the atoms are connected and the bond type. For example, the molecular formula C_4H_{10} can be represented by two different structures: *n*-butane (4) and isobutane (2-methylpropane) (5).

$$\overset{\displaystyle H\ \ H\ \ H\ \ H}{\underset{\displaystyle H\ \ H\ \ H\ \ H}{H-C-C-C-C-H}} \quad \textit{n-}\text{Butane (4)} \qquad\qquad \overset{\displaystyle H\ \ H\ \ H}{\underset{\displaystyle H\ \ \ \ H}{H-C-C-C-H}} \quad \begin{array}{l}\textbf{Isobutane (5)}\\ \textbf{(2-methylpropane)}\end{array}$$

$$H-C-H$$
$$H$$

Consider also the molecular formula C_2H_6O. There are two structures which correspond to this formula: dimethyl ether (6) and ethanol (ethyl alcohol) (7).

H−C−O−C−H

Dimethyl ether (6)

H−C−C−O−H

Ethanol (Ethyl alcohol) (7)

In the pairs above, each structural formula represents a different compound. Each compound has it own unique set of physical and chemical properties. Compounds with the same molecular formula but with different structural formulas are called *isomers*.

The three-dimensional character of molecules is expressed by its *stereochemistry*. By looking at the stereochemistry of a molecule, the spatial relationships between atoms on one carbon and the atoms on an adjacent carbon can be examined. Since rotation can occur around carbon-carbon single bonds in open chain molecules, the atoms on adjacent carbons can assume different spatial relationships with respect to each other. The different arrangements that atoms can assume as a result of a rotation about a single bond are called *conformations*. A specific conformation is called a *conformer*. While individual isomers can be isolated, conformers cannot since interconversion, by rotation, is too rapid.

Conformers may be represented by the use of two conventions as shown in Fig. 1.1.

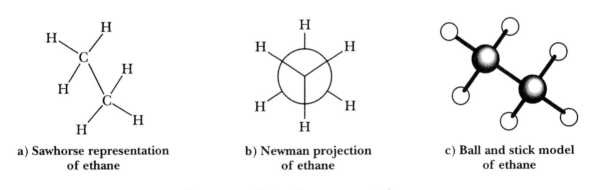

**a) Sawhorse representation
of ethane**　　　**b) Newman projection
of ethane**　　　**c) Ball and stick model
of ethane**

Figure 1.1 Molecular representations.

The *sawhorse* views the carbon-carbon bond at an angle, and by showing all the bonds and atoms, shows their spatial arrangements. The *Newman projection* provides a view along a carbon-carbon bond by sighting directly along the carbon-carbon bond. The near carbon is represented by a circle and bonds attached to it are represented by lines going to the center of the circle; the carbon behind is not visible (since it is blocked by the near carbon), but the bonds attached to it are partially visible and are represented by lines going to the edge of the circle. With Newman projections, rotations show the spatial relationships of atoms on adjacent carbons easily. Two conformers that represent extremes are shown in Fig. 1.2.

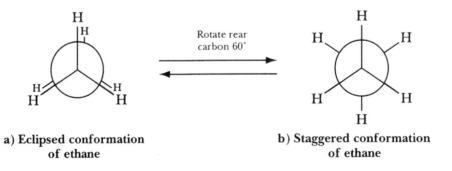

**a) Eclipsed conformation
of ethane**　　　**b) Staggered conformation
of ethane**

Figure 1.2 Two conformers of ethane.

The *eclipsed* conformation has the bonds (and the atoms) on the adjacent carbons as close as possible. The *staggered* conformation has the bonds (and the atoms) on adjacent carbons as far as possible. One conformation can interconvert into the other by rotation around the carbon-carbon bond axis.

The three-dimensional character of molecular structure is shown through molecular model building. With molecular models the number and types of bonds between atoms and the spatial arrangements of the atoms can be visualized for the molecules. This allows comparison of isomers and of conformers for a given set of compounds.

OBJECTIVES

1. To use models to visualize structure in organic molecules.
2. To build and compare isomers having a given molecular formula.
3. To explore the three-dimensional character of organic molecules.

PROCEDURE

Obtain from the laboratory instructor a set of ball and stick molecular models. The set contains the following parts (other colored spheres may be substituted as available):
   a. 2 Black spheres representing *Carbon*; this tetracovalent element has four holes;
   b. 6 Yellow spheres representing *Hydrogen*; this monovalent element has one hole;
   c. 2 Colored spheres representing the *halogen Chlorine*; this monovalent element has one hole;
   d. 1 Blue sphere representing *Oxygen*; this divalent element has two holes;
   e. 8 Sticks to represent bonds.

1. With your models construct the molecule methane. Methane is a simple hydrocarbon consisting of one carbon and four hydrogens. After you put the model together, answer the questions below in the appropriate space on the Report Sheet.
   a. With the model resting so that three hydrogens are on the desk, examine the structure. Move the structure so that a different set of three hydrogens are on the desk each time. Is there any difference between the way that the two structures look (1a)?
   b. Does the term "equivalent" adequately describe the four hydrogens of methane (1b)?
   c. Tilt the model so that only two hydrogens are in contact with the desk and imagine pressing the model flat onto the desk top. Draw the way in which the methane molecule would look in two dimensional space (1c). This is the usual way that three dimensional structures are written.
   d. Using a protractor, measure the angle $H-C-H$ on the model (1d).

2. Replace one of the hydrogens of the methane model with a colored sphere which represents the halogen chlorine. The new model is methyl chloride, CH_3Cl. Position the model so that the three hydrogens are on the desk.
   a. Grasp the atom representing chlorine and tilt it to the right, keeping two hydrogens on the desk. Write the structure of the projection on the Report Sheet (2a).
   b. Return the model to its original position and then tilt, as before, but this time to the left. Write this projection on the Report Sheet (2b).
   c. While the projection of the molecule changes, does the structure of methyl chloride change (2c)?

3. Now replace a second hydrogen with another chlorine sphere. The new molecule is dichloro-methane, CH_2Cl_2.
   a. Examine the model as you twist and turn it in space. Are the projections given below isomers of the molecule CH_2Cl_2 or representations of the same structure only seen differently in three dimensions (3a)?

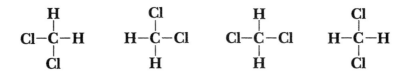

4. Construct the molecule ethane, C_2H_6. Note that you can make ethane from the methane model by removing a hydrogen and replacing the hydrogen with a methyl group, $-CH_3$.
   a. Write the structural formula for ethane (4a).
   b. Are all the hydrogens attached to the carbon atoms equivalent (4b)?
   c. Draw a sawhorse representation of ethane. Draw a staggered and an eclipsed Newman projection of ethane (4c).
   d. Replace any hydrogen in your model with chlorine. Write the structure of the molecule ethyl chloride, C_2H_5Cl (4d).
   e. Twist and turn your model. Draw two Newman projections of the ethyl chloride molecule (4e).
   f. Do the projections that you drew represent different isomers or conformers of the same compound (4f)?

5. Dichloroethane, $C_2H_2Cl_2$
   a. In your molecule of ethyl chloride, if you choose to remove another hydrogen, note that you now have a choice among the hydrogens. You can either remove a hydrogen from the carbon to which the chlorine is attached, or you can remove a hydrogen from the carbon that has only hydrogens attached. First remove the hydrogen from the carbon that has the chlorine attached and replace it with a second chlorine. Write its structure on the Report Sheet (5a).
   b. Compare this structure to the model which would result from removal of a hydrogen from the other carbon and its replacement by chlorine. Write its structure (5b) and compare it to the previous example. One isomer is 1,1-dichloroethane; the other is 1,2-dichloro-ethane. Label the structures drawn on the Report Sheet with the correct name.
   c. Are all the hydrogens of ethyl chloride equivalent? Are some of the hydrogens equivalent? Label those hydrogens which are equivalent to each other (5c)?

6. Butane
   a. Butane has the formula C_4H_{10}. With help from a partner, construct a model of *n*-butane by connecting the four carbons in a series (C–C–C–C) and then adding the needed hydrogens. First orient the model in such a way that the carbons appear as a straight line. Next tilt the model so that the carbons appear as a zig-zag line. Then twist around any of the C–C bonds so that a part of the chain is at an angle to the remainder. Draw each of these structures in the space on the Report Sheet (6a). Note that the structures you draw are for the same molecule but represent only a different orientation and projection.
   b. Sight along the carbon-carbon bond of C_2 and C_3 on the butane chain:

$$\overset{1}{CH_3}-\overset{2}{CH_2}-\overset{3}{CH_2}-\overset{4}{CH_3}.$$

Draw a staggered Newman projection. Rotate the C_2 carbon clockwise by 60°; draw the eclipsed Newman projection. Again rotate the C_2 carbon clockwise by 60°; draw the Newman projection. Is the last projection staggered or eclipsed? (6b). Continue rotation of the C_2 carbon clockwise by 60° increments and observe the changes that take place.

c. Examine the structure of *n*-butane for equivalent hydrogens. In the space on the Report Sheet (6c) redraw the structure of *n*-butane and label those hydrogens which are equivalent to each other. On the basis of this examination, predict how many monochlorobutane isomers (C_4H_9Cl) that could be obtained (6d). Test your prediction by replacement of hydrogen by chlorine on the models. Draw the structures of these isomers (6e).

d. Reconstruct the butane system. First form a three carbon chain, then connect the fourth carbon to the center carbon of the three carbon chain. Add the necessary hydrogens. Draw the structure of isobutane (6f). Can any manipulation of the model, by twisting or turning of the model or by rotation of any of the bonds, give you the *n*-butane system? If these two, *n*-butane and isobutane (2-methylpropane), are *isomers*, then how may we recognize that any two structures are isomers (6g)?

e. Examine the structure of isobutane for equivalent hydrogens. In the space on the Report Sheet (6h), redraw the structure of isobutane and label the equivalent hydrogens. Predict how many monochloroisobutanes could be formed (6i) and test your prediction by replacement of hydrogen by chlorine on the model. Draw the structures of these isomers (6j).

7. C_4H_6O *4 isomers*

a. There are two isomers with the molecular formula C_2H_6O, ethanol (ethyl alcohol) and dimethyl ether. With your partner, construct both of these isomers. Draw these isomers on the Report Sheet (7a) and name each one.

b. Manipulate each model. Can either be turned into the another by a simple twist or turn (7b)?

c. For each compound, label those hydrogens which are equivalent. How many sets of equivalent hydrogens are there in ethanol (ethyl alcohol) and dimethyl ether (7c)?

8. Optional: Butenes

a. If springs are available for the construction of double bonds, construct 2-butene, $CH_3-CH=CH-CH_3$. There are two isomers for compounds of this formulation: the isomer with the 2 CH_3 groups on the same side of the double bond, *cis*-2-butene; the isomer with the 2 CH_3 groups on opposite sides of the double bond, *trans*-2-butene. Draw these two structures on the Report Sheet (8a).

b. Can you twist, turn or rotate one model into the other? Explain (8b).

c. How many bonds are connected to any single carbon of these structures (8c)?

d. With the protractor measure the $C-C=C$ angle (8d).

e. Construct methylpropene, $CH_3-\underset{\underset{\displaystyle CH_3}{|}}{C}=CH_2$. Can you have a *cis*- or a *trans*- isomer in this

system (8e)?

9. Optional: Butynes

a. If springs are available for the construction of triple bonds, construct 2-butyne, $CH_3-C\equiv C-CH_3$. Can you have a *cis*- or a *trans*- isomer in this system (9a)?

b. With the protractor measure the $C-C\equiv C$ angle (9b).

c. Construct a second butyne with your molecular models and springs. How does this isomer differ from the one in (a) above (9c)?

CHEMICALS AND EQUIPMENT

1. Molecular models (you may substitute other available colors for the spheres):
    2 Black spheres
    6 Yellow spheres
    2 Colored spheres (e.g. green)
    1 Blue sphere
    8 Sticks
2. Protractor
3. Optional: 3 springs

4. Ethane and ethyl chloride

   a.

   b.

   c.

   d.

   e.

   f.

5. Dichloroethane

   a.

   b.

   c.

6. Butane

    a.

    b.

    c.

    d.

    e.

    f.

    g.

    h.

    i.

    j.

7. C_2H_6O

   a.

   b.

   c. Ethyl alcohol has _____ set(s) of equivalent hydrogens.

      Dimethyl ether has _____ set(s) of equivalent hydrogens.

8. Butenes

   a.

   b.

   c.

   d. C-C=C angle

   e.

9. Butynes

   a.

   b.

   c.

POST-LAB QUESTIONS

1. There are three (3) isomers of formula, C_3H_8O. Write structural formulas for these compounds.

2. Draw the structure of propane and identify equivalent hydrogens. Identify equivalent sets by letters, e.g., H_a, H_b, etc.

3. Draw the structural formulas for all the isomers of C_4H_9Br.

4. Draw the structural formulas for the four (4) isomers of the butenes, C_4H_8, which are alkenes. Label *cis-* and *trans-* isomers.

5. Draw a staggered and an eclipsed conformer for propane, $\overset{1}{CH_3}-\overset{2}{CH_2}-\overset{3}{CH_3}$, sighting along the C_1-C_2 bond.

EXPERIMENT

2

The Separation of the Components of a Mixture

BACKGROUND

Mixtures are a common occurrence in our lives and are not unique to chemistry. The beverages we drink each morning, the fuel we use in our automobiles and the ground we walk on are mixtures. Very few materials we encounter are pure, homogeneous substances. Any material made up of two or more substances that are not chemically combined is a mixture.

The isolation of pure components of a mixture requires the separation of one component from another. Chemists have developed techniques for the separation or isolation of pure substances from a mixture. These methods take advantage of the differences in physical properties of the components. The techniques to be demonstrated in this laboratory are the following:

1. *Sublimation.* This involves heating a solid until it passes directly from the solid phase into the gaseous phase. The reverse process, the vapor going back to the solid phase without a liquid state in between, is called condensation or deposition. Some solids which sublime are iodine, caffeine, naphthalene and *para*-dichlorobenzene (moth balls).

2. *Extraction.* This uses a suitable solvent to selectively dissolve one component of a mixture. With this technique a soluble compound can be separated from an insoluble compound.

3. *Decantation.* This separates a liquid from an insoluble solid sediment by carefully pouring the liquid from the solid without disturbing the solid (Fig. 2.1).

Figure 2.1 Decantation.

4. *Filtration.* This separates a solid from a liquid through the use of a porous material, a filter. Paper, charcoal, or sand can serve as a filter. These materials allow the liquid to pass through, but not the solid (see Fig. 2.4 in the Procedure section).

5. *Evaporation.* This is the process of heating a mixture in order to drive off, in the form of vapor, a volatile liquid, so as to make the remaining component dry.

The mixture that will be separated in this laboratory experiment contains three components: naphthalene ($C_{10}H_8$), common table salt (NaCl) and sea sand (SiO_2). The separation will be done according to the scheme in Fig. 2.2 by

1. heating the mixture to sublime the naphthalene;
2. dissolving the table salt with water to extract; and
3. evaporating water to recover dry NaCl and sand.

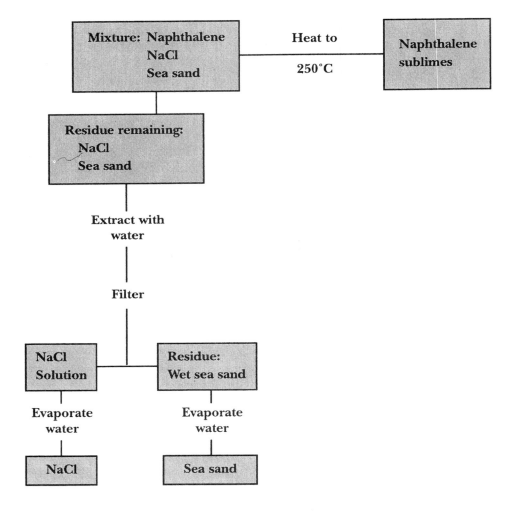

Figure 2.2 Separation scheme.

OBJECTIVES

1. To demonstrate the separation of a mixture
2. To examine some techniques for separation using physical methods.

PROCEDURE

1. Obtain a clean, dry 150-mL beaker and carefully weigh to the nearest 0.001 g; record this weight for beaker 1 on the Report Sheet (1). Obtain from your instructor a sample of the unknown mixture; use a mortar and pestle to grind the mixture into a fine powder. With the beaker still on the balance, carefully transfer approximately 2 g of the unknown mixture into the beaker. Record the weight of the beaker with the contents to the nearest 0.001 g (2). Calculate the exact sample weight by subtraction (3).

2. Place an evaporating dish on top of the beaker containing the mixture. Place the beaker and evaporating dish on a wire gauze with an iron ring and ring stand assembly as shown in Fig. 2.3. Place ice in the evaporating dish being careful not to get any water on the underside of the evaporating dish or inside the beaker.

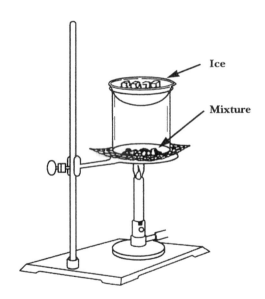

Figure 2.3 Assembly for sublimation.

3. Carefully heat the beaker with a Bunsen burner, increasing the intensity of the flame until fumes appear in the beaker. Solid should collect on the underside of the evaporating dish. After 10 min. of heating, remove the flame from the beaker. Carefully remove the evaporating dish from the beaker and collect the solid by scraping it off the dish with a spatula. Drain away any water from the evaporating dish and add ice to it if necessary. Stir the contents of the beaker with a glass rod. Return the evaporating dish to the beaker and apply the heat again. Continue heating and scraping off solid until no more solid collects. Discard the naphthalene into a special container provided by your instructor.

4. Allow the beaker to cool until it reaches room temperature and then weigh the beaker with the contained solid (4). Calculate the weight of the naphthalene that sublimed (5).

5. Add 25 mL of distilled water to the solid in the beaker; heat and stir for 5 min.

6. Weigh a second clean, dry 150-mL beaker to the nearest 0.001 g (6).

7. Assemble the apparatus for gravity filtration as shown in Fig. 2.4.

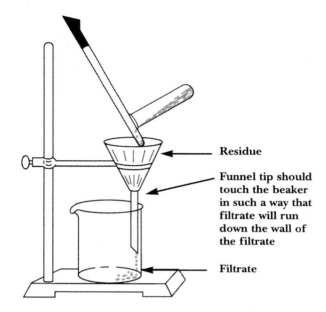

Figure 2.4 Gravity filtration.

8. Fold a piece of filter paper following the technique shown in Fig. 2.5.

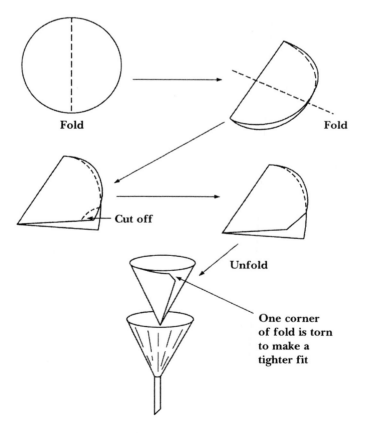

Figure 2.5 The steps for folding a filter paper for a gravity filtration.

9. Wet the filter paper with water and adjust the paper so that it lies flat on the glass of the funnel.

10. Position the second weighed beaker under the funnel.

11. First decant most of the liquid into beaker 2. Then pour the mixture through the filter by carefully transferring the wet solid into the funnel with a spatula or rubber policeman. Collect all the liquid (called the filtrate) in beaker 2.

12. Rinse beaker 1 with 5 - 10 mL of water, pour over the residue in the funnel, and add the liquid to the filtrate; repeat with an additional 5 - 10 mL of water.

13. Place beaker 2 and its contents on a wire gauze with an iron ring and ring stand assembly as shown in Fig. 2.6 A.

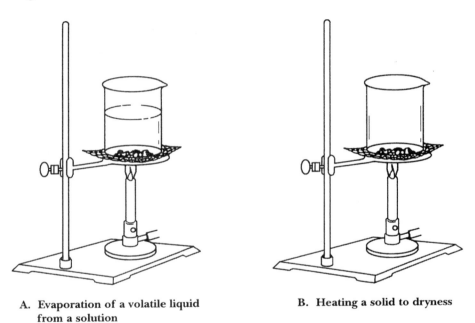

A. Evaporation of a volatile liquid   B. Heating a solid to dryness
   from a solution

Figure 2.6 Assembly for evaporation.

Add boiling chips, two or three are enough, to the solution and begin to heat gently with a Bunsen burner. Control the flame in order to prevent boiling over. As the volume of liquid is reduced, solid sodium chloride will appear. Reduce the flame to avoid bumping of the solution and spattering of the solid. When all of the liquid is gone, cool the beaker to room temperature. With a spatula or forceps, remove the boiling stones. Weigh the beaker and the solid residue to the nearest 0.001 g (7). Calculate the weight of the recovered NaCl by subtraction (8).

14. Carefully weigh a third clean, dry 150-mL beaker to the nearest 0.001 g (9). Transfer the sand from the filter paper to beaker 3. Heat the sand to dryness in the beaker with a Bunsen burner using the ring stand and assembly shown in Fig. 2.6 B (or use an oven at T = 90 − 100°C if available); heat carefully to avoid spattering; when dry the sand should be freely flowing. Allow the sand to cool to room temperature. Weigh the beaker and the sand to the nearest 0.001 g (10). Calculate the weight of the recovered sand by subtraction (11).

15. Calculate

    a. Percentage yield using the formula:

    $$\% \text{ yield} = \frac{\text{grams of solid recovered}}{\text{grams of initial sample}} \times 100$$

    b. Percentage of each component in the mixture by using the formula:

    $$\% \text{ component} = \frac{\text{grams of component isolated}}{\text{grams of initial sample}} \times 100$$

EXAMPLE

A student isolated from a sample of 1.132 g the following:

> 0.170 g of naphthalene
> 0.443 g of NaCl
> 0.499 g of sand
> _____
> 1.112 g solid recovered

The student calculated the percentage yield and percentage of each component as follows (to correct number of significant figures):

$$\% \text{ yield} = \frac{1.112}{1.132} \times 100 = 98.23\%$$

$$\% \text{ C}_{10}\text{H}_8 = \frac{0.170}{1.132} \times 100 = 15.0\%$$

$$\% \text{ NaCl} = \frac{0.443}{1.132} \times 100 = 39.1\%$$

$$\% \text{ sand} = \frac{0.499}{1.132} \times 100 = 44.1\%$$

CHEMICALS AND EQUIPMENT

1. Unknown mixture
2. Balances
3. Boiling stones
4. Evaporating dish, porcelain, 6 cm
5. Filter paper, 15 cm
6. Mortar and pestle
7. Oven (if available)
8. Ring stands (2)
9. Rubber policeman
10. Spatula

EXPERIMENT 2

NAME _____ SECTION _____ DATE _____

PARTNER _____ GRADE _____

PRE-LAB QUESTIONS

1. List five (5) methods that could be used to separate the components found in a mixture.

2. Can any of the methods listed above be used to separate the elements found in a compound? Explain.

3. How do decantation and filtration differ?

4. Define sublimation.

EXPERIMENT 2

NAME _____ **SECTION** _____ **DATE** _____

PARTNER _____ **GRADE** _____

REPORT SHEET

1. Weight of beaker 1   _____ g

2. Weight of beaker 1 and mixture   _____ g

3. Weight of mixture: (2) − (1)   _____ g

4. Weight of beaker 1 and solid after sublimation   _____ g

5. Weight of naphthalene: (2) − (4)   _____ g

6. Weight of beaker 2   _____ g

7. Weight of beaker 2 and NaCl   _____ g

8. Weight of NaCl: (7) − (6)   _____ g

9. Weight of beaker 3   _____ g

10. Weight of beaker 3 and sand   _____ g

11. Weight of sand: (10) − (9)   _____ g

Calculations

12. Weight of recovered solids: (5) + (8) + (11)   _____ g

13. Percentage yield (percentage of solids recovered):

    [(12)/(3)] × 100   _____ %

14. Percentage of naphthalene: [(5)/(3)] × 100   _____ %

15. Percentage of NaCl: [(8)/(3)] × 100   _____ %

16. Percentage of sand: [(11)/(3)] × 100   _____ %

POST-LAB QUESTIONS

1. A student started with a mixture weighing 1.356 g. The student recovered 1.543 g. Assuming that all calculations were done correctly, what was the most likely source of error in the experiment?

2. What error would result if the sand in the experiment was not rinsed?

3. *para*-Dichlorobenzene is sold commercially as moth balls. Why can it be used successfully in a closed garment bag to prevent damage to clothes by moth larvae?

4. How can a mixture of sugar and water be separated?

5. A 70 kg man has 19.8 kg of his body weight as fat. What is the percentage of fat in the man?

6. Dry cleaners remove oil and grease spots from clothing by using an organic solvent called perchloroethylene. What method of separation did the cleaner use?

7. What method of separation was used in order to prepare a pot of coffee from the ground coffee beans or a pot of tea from the tea leaves?

8. Ice kept in an open container below 0°C will eventually disappear. How does this happen?

EXPERIMENT 3

Resolution of a Mixture by Distillation

BACKGROUND

Simple Distillation

One of the most common methods of purifying a liquid is that of distillation. It is a very simple method: a liquid in a container is brought to a boil; the liquid vaporizes; the vapor is cooled in a condenser and returns to the liquid state; the liquid is collected in a separate container.

Everyone has had an opportunity to heat water to a boil in a vessel. As heat is applied, water molecules increase in their kinetic energy. Some molecules acquire sufficient energy to escape from the liquid phase and enter into the vapor phase. The vapor above the liquid exerts a pressure, called the vapor pressure. As more and more molecules obtain enough energy to escape into the vapor phase, the pressure exerted by these molecules in the vapor phase increases. Eventually the vapor pressure equals the pressure exerted externally on the liquid (this external pressure usually is caused by the atmosphere). When this condition is met, boiling occurs and the temperature at which this occurs is the boiling point.

Normal distillation, procedures carried out at atmospheric pressure, require "normal" boiling points. However, when boiling takes place in a closed system, it is possible to change the boiling point of the liquid by changing the pressure in the closed system. If the external pressure is reduced, usually by using a vacuum pump or water aspirator, the boiling point of the liquid is reduced. Thus, heat sensitive liquids which decompose when boiled at atmospheric pressure, distill with minimum decomposition at reduced pressure and temperature. The relation of

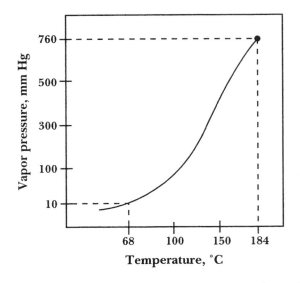

Figure 3.1 Temperature-vapor pressure curve for aniline.

temperature to vapor pressure for aniline can be shown by the curve in Fig. 3.1. The organic liquid aniline, $C_6H_5NH_2$, can be distilled at 184°C (760 mm Hg) or at 68°C (10 mm Hg).

In a distillation, the process described is carried out in a system such as illustrated in Fig. 3.2. The liquid in the boiling flask is heated to a boil and the vapor rises through tubing. The vapor then travels into a tube cooled by water, a condenser, where the vapor returns to the liquid state. If the mixture has a low boiling component (a volatile substance with a high vapor pressure), it will distill over first and can be collected. Higher boiling and nonvolatile components (substances with low vapor pressure) remain in the boiling flask. Only by applying more heat could the higher boiling component be distilled. Nonvolatile substances would not distill.

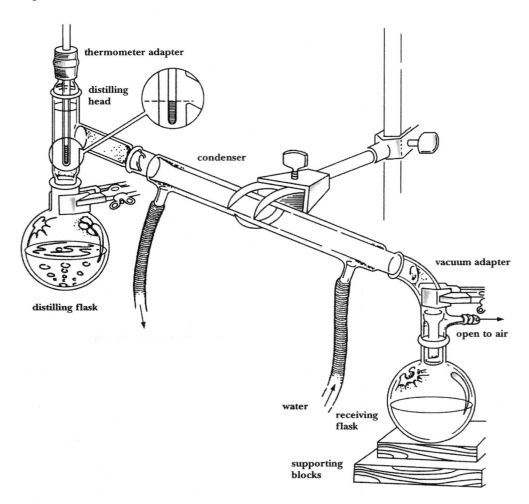

Figure 3.2 A simple distillation apparatus.

Consider a mixture of two miscible, volatile liquids that have different boiling points. Fig. 3.3 is a typical diagram of the liquid-vapor relationship of a two-component mixture. This type of diagram is referred to as a *phase diagram.*

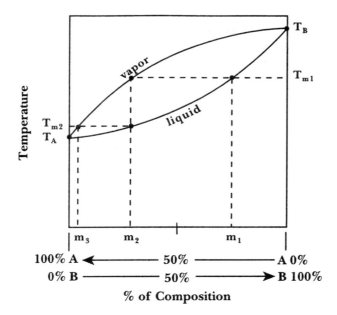

Figure 3.3 Phase diagram of a two component mixture.

T_A is the boiling point of pure A; T_B is the boiling point of pure B. The two boiling points are connected by two curves: the lower curve represents the composition of the liquid mixture at various boiling points; the upper curve represents the composition of the vapor in equilibrium with the liquid at those temperatures. Notice that the boiling points rise as the mixture increases in the higher boiling component.

Suppose there is a liquid mixture composed of 80% B and 20% A; this is m_1 on the diagram in Fig. 3.3. As the liquid is heated, the temperature will rise (follow the dotted lines) until it reaches the boiling point, T_{m1}. Liquid vaporizes and the vapor at temperature T_{m1} has the composition m_2. Note that the composition of the vapor is *richer in* A, the lower boiling, more volatile component; condensation of this vapor gives a liquid of composition m_2, a liquid richer in A than the original liquid m_1. Collection of this liquid gives a first fraction and shows that component A has been partially separated from component B.

Since component A has been selectively removed from the original liquid mixture, the remaining liquid is richer in component B. This results in a higher boiling point for the mixture; the boiling point would follow the curve and rise toward T_B. In turn as this new mixture boils, the vapor in equilibrium changes; the vapor has less of component A and more of component B. Collection of this mixture on condensation gives a *second fraction*. Eventually the material collected in the *last fraction* contains mostly the higher boiling, less volatile component B. A *simple distillation* using the apparatus shown in Fig. 3.2 brings about the separations described.

Further purifications can be done. If the liquid from the first fraction, m_2, is heated to its boiling point, T_{m2}, vapor of composition m_3 is obtained; again note that the vapor is *richer in* A, as is the condensed liquid. If this process were continued (vaporization, condensation, etc.), a pure sample of A results. In the same way, the higher boiling fractions can be redistilled until pure B results.

In general, two miscible, volatile liquids whose boiling points differ by a minimum of 25°C can be separated to some degree by a simple distillation.

Gas Chromatography

Gas chromatography (GC) or vapor phase chromatography (VPC) is a useful technique for separating and analyzing organic compounds. The method works well when the compounds are volatile and do not decompose when vaporized. GC can rapidly 1) **test the purity of a substance, 2) show the number of components in a mixture, 3) give the relative percentage composition of the components in a mixture, and 4) determine the identity of a component, in some cases.**
A schematic diagram of the apparatus is shown in Fig. 3.4.

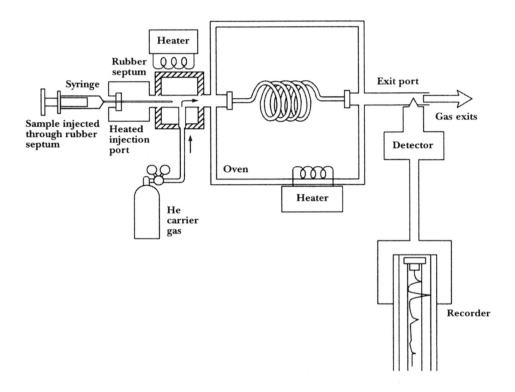

Figure 3.4 The components of a gas chromatograph.

The operation of the instrument is as follows. The injection port, oven, and detector are heated to the temperature required for the experiment. A small sample of the material to be analyzed is injected with a syringe through a rubber septum into the heated injection port. The sample is vaporized and is carried into a packed column by an inert carrier gas, usually helium. [The column is a tube of varying length (0.5 m to 30 m in length, 0.5 mm to 6 mm in diameter) packed with a porous solid that is coated with a thin layer of a high-boiling liquid; the column is in a tempera-ture controlled oven.] The sample is carried through the column and is separated into its components. (Variables which influence the rate a given component passes through the column include: *boiling point, polarity, solubility of the component in the high-boiling liquid, column temperature, carrier gas flow rate*.) As each component exits (or elutes) from the column, it enters the detector; this is an electronic device that generates a signal when a component is detected. The signal is transmitted to a strip chart recorder which provides a record of the analysis.
A typical trace of a two-component mixture is shown in Fig. 3.5.

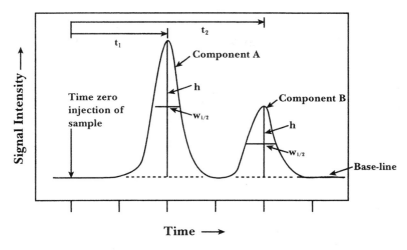

Figure 3.5 Gas chromatograph of a two-component system.

Since there are two peaks in the trace, the sample is a mixture of two components (a pure sample would show only a single peak). The time it took from the point of injection for each component to elute from the column is the *retention time*: t_1 for component A; t_2 for component B. The retention time, when determined under identical, controlled conditions, can be used to identify a component; i.e. a direct comparison of an unknown component's retention time with a known compound should be the same under the same conditions. Finally, *the area under each curve approximates the relative percentage composition of each component in the mixture.* (One method of determining relative area is by triangulation. The area of component A, A_A, is found by multiplying the height, h, of its peak above the base-line by the width of its peak at half-height, $w_{1/2}$: $A_A = (h \times w_{1/2})_A$. Area of component B, A_B, is found the same way: $A_B = (h \times w_{1/2})_B$. The total area contributed by all the component peaks is $A_T = A_A + A_B$. The relative percent of any component in the mixture is found by dividing the individual area by the total area: $\%A = A_A/A_T$; $\%B = A_B/A_T$.)

There are many different gas chromatographs sold commercially. They each differ in cost, complexity and ability to separate and detect components in a mixture. However, they all work essentially as described. Your instructor will demonstrate the instrument available in your laboratory and will explain its unique characteristics.

OBJECTIVES

1. To use distillation to *separate* a mixture.
2. To show that distillation can *purify* a liquid.
3. To use gas chromatography to *analyze* volatile mixtures.

PROCEDURE

This experiment consists of three parts. Consult your instructor as to which parts to complete.

Part A: A salt-water mixture will be separated by a distillation. The volatile water will be separated from the nonvolatile salt, sodium chloride, NaCl. The purity of the collected distilled water will be demonstrated by chemical tests specific for sodium ions, Na^+, and chloride ions, Cl^- .

Part B: You will distill a mixture of cyclohexane and toluene and collect four (4) fractions. A simple distillation will be carried out.

Part C: Each fraction collected from Part B will be analyzed by gas chromatography.

Part A. A salt-water mixture.

1. Assemble an apparatus as illustrated in Fig. 3.2. A kit containing the necessary glassware can be obtained from your instructor. The glassware contains standard taper joints which allow for quick assembly and disassembly. Before fitting the pieces together, apply a light coating of silicone grease to each joint to prevent the joints from sticking.

2. Use 100-mL round bottom flasks for the boiling flask and the receiving flask. Fill the boiling flask with 50 mL of the prepared salt-water mixture. Add two boiling chips to the boiling flask to ensure smooth boiling of the mixture and to prevent bumping. Be sure that the rubber tubing to the condenser enters the lower opening and empties out the upper opening. Turn on the water faucet and allow the water to fill the jacket of the condenser slowly so as not to trap air. Take care not to provide too much flow, otherwise the hoses will disconnect from the condenser. Adjust the mercury bulb of the thermometer to be below the junction of the condenser with the distillation head. Be sure the vacuum adapter is open to the air.

3. Gently heat the boiling flask with a heating mantle by adjusting the Variac. (Your instructor may have other heat sources available and will direct you in their use.) Eventually the liquid will boil, vapors will rise and enter the condenser, and the vapors will condense and be collected in the receiving flask.

4. Discard the first one mL of water collected. Record the temperature of the vapors as soon as the one mL of water has been collected. Continue collection of the distilled water until approximately one-half of the mixture has distilled. Record the temperature of the vapors at this point. Turn off the heat and allow the system to return to room temperature.

5. The distilled water in the receiving flask and the liquid in the boiling flask will be tested.

6. Place in separate clean, dry test tubes (100 × 13 mm) 2 mL of distilled water and 2 mL of the residue liquid from the boiling flask. Add to each sample 5 drops of silver nitrate solution. Look for the appearance of a white precipitate. Record your observations. Silver ions combine with chloride ions to form a white precipitate of silver chloride.

$$Ag^+ + Cl^- \rightarrow AgCl_{(s)} \text{ (White precipitate)}$$

7. Place in separate clean, dry test tubes (100 × 13 mm) 2 mL of distilled water and 2 mL of the residue liquid from the boiling flask. Obtain from your instructor a clean nickel wire. (**CAUTION!** Do the following tests **in the hood**.) Dip the wire into concentrated nitric acid and hold the wire in a Bunsen burner flame until the yellow color in the flame disappears. Dip the wire into the distilled water sample and put the wire into the Bunsen burner flame. Record the color of the flame. Repeat the above procedure, cleaning the wire, dipping the wire into the liquid from the boiling flask and observing the color and the intensity of the Bunsen burner flame. Record your observations. Sodium ions produce a bright yellow flame with a Bunsen burner.

8. Make sure you wipe the grease from the joints before washing the glassware used in the distillation.

Part B. Cyclohexane and toluene mixture.

1. Assemble an appparatus as illustrated in Fig. 3.2. A kit containing the necessary glassware can be obtained from your instructor. The glassware contains standard taper joints which allow for quick assembly and disassembly. Before fitting the pieces together, apply a light coating of silicone grease to each joint to prevent the joints from sticking.

2. Use a 25-mL round bottom flask for the boiling flask and 10-mL round bottom flasks for the receiving flasks. Add to the boiling flasks 5 mL of cyclohexane and 5 mL of toluene; add two (2) boiling chips to the boiling flask, before heating, to ensure smooth boiling of the mixture and to prevent bumping.

*Use **NO** flames. Cyclohexane and toluene are flammable; Bunsen burners cannot be used for heating. Do not add boiling stones to hot liquids.*

Be sure that all connecting joints are snug and do not leak. Connect the rubber tubing to the condenser so that water enters the lower opening and empties out the upper opening. Turn on the water faucet and allow the water to fill the jacket of the condenser slowly so as not to trap air. Take care not to provide too much flow, otherwise the hoses will disconnect from the condenser. Adjust the bulb of the thermometer to be below the junction of the condenser with the distilling head.

3. Gently heat the boiling flask with a heating mantle by adjusting the Variac. Eventually the liquid will boil, vapors will rise and enter the condenser, and the vapors will condense and collect in the receiving flask.

4. Collect four (4) fractions in the following ranges: 1) 75 - 85°C; 2) 85 - 95°C; 3) 95 - 105°C; 4) 105 - 111°C.

Do not heat the boiling flask to dryness; if peroxides are present, an explosion may occur.

Screw top vials

Each of these fractions will be collected in a separate ~~10-mL round bottom flask~~. Measure the volume for each of the liquid condensate fractions in a graduate cylinder and record on the Report Sheet. Store each fraction in a labeled (by fraction number), clean, dry, screw cap vial; line the cap with aluminum foil. These fractions, at the option of your instructor, will be used in Part C.

5. Turn off the heat and water and allow the system to return to room temperature. Make sure you wipe the grease from the joints before washing the glassware used in the distillation.

Part C. Gas Chromatography (GC).

1. The gas chromatograph will be set-up for you by the instructor. The column will be installed, all the temperatures will be set and the carrier gas will be set to the proper flow rate. Follow the instructions for operating the chromatograph, adjusting the recorder and injecting the sample. Record on the Report Sheet the conditions of the chromatograph for your separation.

2. Your instructor will have available a gas chromatogram for pure cyclohexane and pure toluene at the conditions you will use. Use this for comparison to your gas chromatograms in order to identify cyclohexane and toluene in your fractions.

3. Inject a small sample (2 - 5 μL), one at a time, from each of fraction 1, fraction 2, fraction 3 and fraction 4 obtained from the simple distillation procedure. Determine the relative areas of each component in each fraction and record the results on the Report Sheet. [On a Hewlett-Packard Model 5890 Chromatograph using an HP-1 crosslinked methyl silicone gum column (30 m $\times$ 0.53 mm $\times$ 2.65 μm film thickness) at a column temperature of 100°C, cyclohexane has a retention time of 1.0 min. and toluene has a retention time of 1.4 min.]

4. Dispose of any cyclohexane and toluene in waste containers provided by your instructor.

CHEMICALS AND EQUIPMENT

1. Clamps
2. Distillation kit
3. Heating mantle
4. Thermometer
5. Variac
6. Boiling chips
7. Nickel wire
8. Concentrated nitric acid, HNO_3
9. Salt-water mixture
10. 0.5 M silver nitrate, $AgNO_3$
11. Silicone grease
12. Cyclohexane
13. Toluene
14. 10 μL syringe
15. Gas chromatograph

EXPERIMENT 3

NAME _____   SECTION _____   DATE _____

PARTNER _____   GRADE _____

PRE-LAB QUESTIONS

1. How do you define the boiling point?

2. In the distillation set-up, what is the purpose of the water-cooled condenser?

3. Sea water is a mixture which has salt, NaCl, dissolved in water. How can these compounds be separated?

4. A student has a mixture of two liquids. Liquid A boils at 112°C and liquid B boils at 145°C. The first drops of liquid to distill will belong to which liquid? Why?

5. What information can be obtained from gas chromatography?

EXPERIMENT 3

NAME _____   SECTION _____ DATE _____

PARTNER _____   GRADE _____

REPORT SHEET

Part A. Salt-water mixture

1. Barometric pressure   _____

2. Boiling point of water at measured pressure   _____

3. Temperature of vapor after collecting 1 mL   _____

4. Temperature of vapor at end of distillation   _____

~~Solution~~	Observation with 0.5 M $AgNO_3$	Color in Flame Test
~~Distilled water~~		
~~Liquid in boiling flask~~		

Part B.   Cyclohexane and toluene mixture

Part C.   Gas Chromatography

From GC Data
Ethanol   *% 1-butanol*

Fraction No.	b. p. Range(°C)	Volume(mL)	~~% Cyclohexane~~	~~% Toluene~~
1	79 ~~75~~ - 85			
2	85 - 95			
3	95 - 105			
4	105 - ~~111~~ 117			

GC Conditions

1. Type of column:

2. Length and diameter of column:

3. Temperature of column:

4. Carrier gas:

5. Flow rate:

6. Retention time of cyclohexane:

7. Retention time of toluene:

POST-LAB QUESTIONS

1. The temperature a student observed during a distillation differed from the literature value for the boiling point of water. What factor or factors may account for the difference?

2. In this experiment, why didn't the distilled water test positively for chloride anion, Cl^-?

3. Hikers at high altitudes in the Alps find that water will not boil at 100°C. Will the boiling point be lower or higher? Explain.

4. What could happen if a distillation were carried out with the vacuum adapter closed to air and all the joints were sealed tightly (see Fig. 3.2)?

5. In the simple distillation of cyclohexane and toluene, what trend was apparent in the compositions of fraction 1 through fraction 4?

6. How can gas chromatography be used in the identification of a compound?

7. A student attempted to distill a mixture of hexane (b. p. 69°C), ethyl acetate (b. p. 77°C) and toluene (b. p. 111°C). A fraction was obtained with a boiling point range of 70 - 90°C. What should the gas chromatogram look like? (Show the curve or curves plotting intensity or height *versus* time.)

8. A student took a sample of liquid for a gas chromatograph. The following curves were obtained:

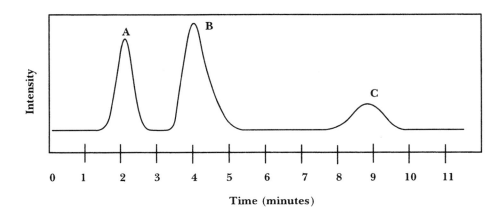

a. The area under peak A was 35 units², peak B 45 units², and peak C 20 units². What percentage of the product mixture is

    A?   B?   C?

b. The student expected three products: 3-methylcyclohexene with a retention time of 2 min., 1-methylcyclohexene with a retention time of 4 min., and 2-methylcyclohexanol with retention time of 9 min. What is the most likely identity of

    A?   B?   C?

Physical Properties of Chemicals: Melting Point, Sublimation, and Boiling Point

BACKGROUND

If you were asked to describe a friend, most likely you would start by identifying particular physical characteristics. You might begin by giving your friend's height, weight, hair color, eye color or facial features. These characteristics would allow you to single out the individual from a group.

Chemicals also possess distinguishing physical properties which enable their identification. In many circumstances a thorough determination of the physical properties of a given chemical can be used for its identification. If faced with an unknown sample, a chemist may compare the physical properties of the unknown to properties of known substances that are tabulated in the chemical literature; if a match can be made, an identification can be assumed (unless chemical evidence suggests otherwise).

The physical properties most commonly listed in handbooks of chemical data are color, crystal form (if a solid), refractive index (if a liquid), density, solubility in various solvents, melting point, sublimation characteristics and boiling point. When a new compound is isolated or synthesized, these properties almost always accompany the report in the literature.

The transition of a substance from a solid to a liquid to a gas, and the reversal, represent physical changes. This means that there is a change in the form or the state of the substance without any alteration in the chemical composition. Water undergoes state changes from ice to liquid water to steam; however, the composition of molecules in all three states remains H_2O.

$$H_2O_{(s)} \;\; \Longleftrightarrow \;\; H_2O_{(l)} \;\; \Longleftrightarrow \;\; H_2O_{(g)}$$

$$\text{Ice} \qquad\qquad \text{Liquid} \qquad\qquad \text{Steam}$$

The *melting* or *freezing point* of a substance refers to the temperature at which the solid and liquid states are in equilibrium. The terms are interchangeable and correspond to the same temperature; how the terms are applied depends upon the state the substance is in originally. The melting point is the temperature at equilibrium when starting in the solid state and going to the liquid state. The freezing point is the temperature at equilibrium when starting in the liquid state and going to the solid state.

Melting points of pure substances occur over a very narrow range and are usually quite sharp. The criteria for purity of a solid is the narrowness of the melting point range and the correspondence to the value found in the literature. Impurities will lower the melting point and cause a broadening of the range. For example, pure benzoic acid has a reported melting point of 122.13°C; benzoic acid with a melting point range of 121-122°C is considered to be quite pure.

The *boiling point* or *condensation point* of a liquid refers to the temperature when its vapor pressure is equal to the external pressure. If a beaker of liquid is brought to a boil in your laboratory, bubbles of vapor form throughout the liquid; these bubbles rise rapidly to the surface, burst and release vapor to the space above the liquid. In this case the liquid is in contact with the atmosphere; the normal boiling point of the liquid will be the temperature when the pressure of the vapor is equal to the atmospheric pressure (1 atm or 760 mm Hg). Should the external pressure vary, so will the boiling point. A liquid will boil at a higher temperature when the external pressure is higher and at a lower temperature when the external pressure is reduced. The change in state from a gas to a liquid represents condensation and is the reverse of boiling. The temperature for this change of state is the same as the boiling temperature, but is concerned with the approach from the gas phase.

Just as a solid has a characteristic melting point, a liquid has a characteristic boiling point. At one atmosphere pure water boils at 100°C, pure ethyl alcohol boils at 78.5°C and pure diethyl ether boils at 34.6°C. The vapor pressure curves shown in Fig. 4.1 illustrate the variation of the vapor pressure of these liquids with temperature. One can use these curves to predict the boiling point at a reduced pressure. For example, diethyl ether has a vapor pressure of 422 mm Hg at 20°C. If the external pressure were reduced to 422 mm Hg, diethyl ether would boil at 20°C.

Sublimation is a process that involves the direct conversion of a solid to a gas, and its reversal, gas to a solid, without passing through the liquid state, when heated. Relatively few solids do this. Some examples are the solid compounds naphthalene (moth balls), caffeine, iodine and solid carbon dioxide (commercial Dry Ice). Sublimation temperatures are not as easily obtained as melting points or boiling points.

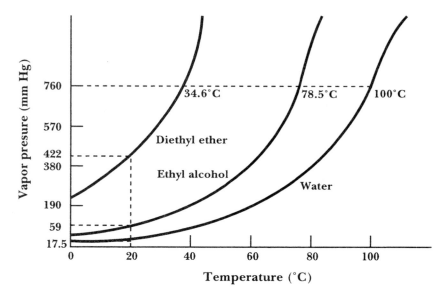

Figure 4.1 Diethyl ether, ethyl alcohol and water vapor pressure curves.

OBJECTIVES

1. To use melting points and boiling points in identifying substances.
2. To use sublimation as a means of purification.

PROCEDURE

Melting point determination

1. Unknowns are provided by the instructor. Obtain approximately 0.1 g of unknown solid and place it on a small watch glass. Record the number of the unknown on the Report Sheet (1). (The instructor will weigh out a 0.1 g sample; take approximately that amount with your spatula.) Carefully crush the solid into a powder with the flat portion of a spatula.

2. Obtain a melting point capillary tube. One end of the tube will be sealed. The tube is packed with solid in the following way:

    Step A.   Press the open end of the capillary tube vertically into the solid sample (Fig. 4.2 A). A small amount of sample will be forced into the open end of the capillary tube.

    Step B.   Invert the capillary tube so that the closed end is pointing toward the bench top. Gently tap the end of the tube against the lab bench top (Fig. 4.2 B). Continue tapping until the solid is forced down to the closed end. A sample depth of 5-10 mm is sufficient.

    Step C.   An alternative method for bringing the solid sample to the closed end uses a piece of glass tubing of approximately 20 to 30 cm. Hold the capillary tube, closed end down, at the top of the glass tubing, held vertically; let the capillary tube drop through the tubing so that it hits the lab bench top. The capillary tube will bounce and bring the solid down. Repeat if necessary (Fig. 4.2 C).

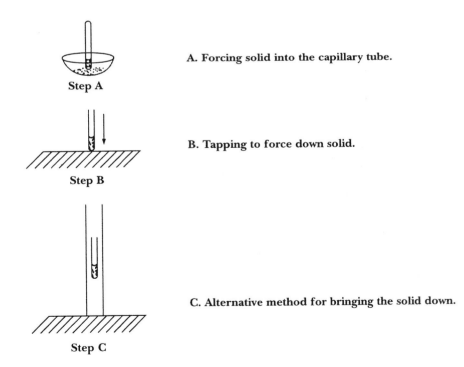

A. Forcing solid into the capillary tube.

B. Tapping to force down solid.

C. Alternative method for bringing the solid down.

Figure 4.2 Packing a capillary tube.

3. The melting point may be determined using either a commercial melting point apparatus or a Thiele tube.
    a.   A commercial melting point apparatus will be demonstrated by your instructor.

b. The use of the Thiele tube is as follows:

   (1) Attach the melting point capillary tube to the thermometer by means of a rubber ring. Align the mercury bulb of the thermometer so that the tip of the melting point capillary containing the solid is next to it (Fig. 4.3).

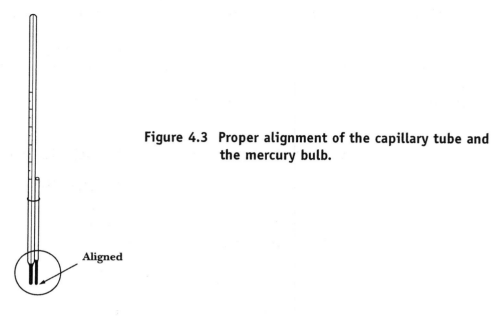

Figure 4.3 Proper alignment of the capillary tube and the mercury bulb.

   (2) Use an extension clamp to support the Thiele tube on a ring stand. Add mineral oil or silicone oil to the Thiele tube, filling to a level above the top of the side arm. Use a thermometer clamp to support the thermometer with the attached melting point capillary tube in the oil. The bulb and capillary tube should be immersed in the oil; keep the rubber ring and open end of the capillary tube out of the oil.

   (3) Heat the arm of the Thiele tube very slowly with a Bunsen burner flame. Use a small flame and gently move the burner along the arm of the Thiele tube (Fig. 4.4).

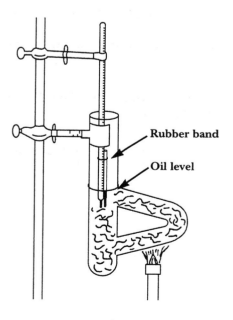

Figure 4.4 Thiele tube apparatus.

(4) You should position yourself so that you can follow the rise of the mercury in the thermometer as well as observe the solid in the capillary tube. Record the temperature when the solid begins to liquify (2) (the solid will appear to shrink). Record the temperature when the solid is completely a liquid (3). Express these readings as a melting point range (4).

(5) Identify the solid by comparing the melting point with the solids listed in Table 4.1 (5).

4. Do as many melting point determinations as your instructor may require. Just remember to use a new melting point capillary tube for each melting point determination.

5. Dispose of the solids as directed by your instructor.

TABLE 4.1 MELTING POINTS OF SELECTED SOLIDS

Solids	Melting Point (°C)
Acetamide	82
Acetanilide	114
Benzophenone	48
Benzoic acid	122
Biphenyl	70
Lauric acid	43
Naphthalene	80
Stearic acid	70

Purification of naphthalene by sublimation

1. Place approximately 0.5 g of impure naphthalene into a 100-mL beaker. (Your instructor will weigh out 0.5 g of sample; with a spatula take an amount which approximates this quantity.)

2. Into the 100-mL beaker place a smaller 50-mL beaker. Fill the smaller halfway with ice cubes or ice chips. Place the assembled beakers on a wire gauze supported by a ring clamp (Fig. 4.5).

3. Using a small Bunsen burner flame, gently heat the bottom of the 100-mL beaker by passing the flame back and forth beneath the beaker.

4. You will see solid flakes of naphthalene collect on the bottom of the 50-mL beaker. When a sufficient amount of solid has collected, turn off the burner.

5. Pour off the ice water from the 50-mL beaker and carefully scrape the flakes of naphthalene onto a piece of filter paper with a spatula.

6. Take the melting point of the pure naphthalene and compare it to the value listed in Table 4.1 (6).

7. Dispose of the crude and pure naphthalene as directed by your instructor.

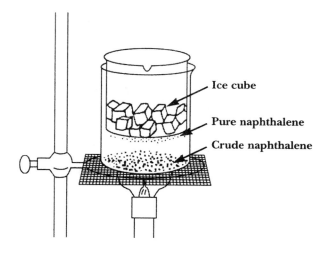

Ice cube

Pure naphthalene

Crude naphthalene

Figure 4.5 Set-up for sublimation of naphthalene.

Boiling point determination

The chemicals used for boiling point determinations are flammable. Be sure all Bunsen burner flames are extinguished before completing this part of the experiment.

1. Obtain from your instructor an unknown liquid and record its number on the Report Sheet (7).

2. Clamp a clean, dry test tube (100 × 13 mm) onto a ring stand. Add to the test tube approximately 3 mL of the unknown liquid and two small boiling chips. Lower the test tube into a 250-mL beaker which contains 100 mL of water and two boiling chips. Adjust the depth of the test tube so that the unknown liquid is below the water level of the water bath (Fig. 4.6).

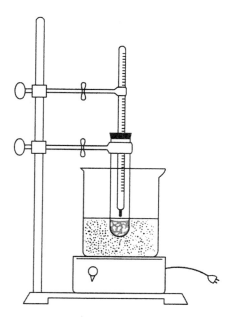

Figure 4.6 Set-up for determining the boiling point.

3. With a thermometer clamp, securely clamp a thermometer and lower it into the test tube through a neoprene adapter. Adjust the thermometer so that it is approximately 1 cm above the surface of the unknown liquid.

4. A piece of aluminum foil can be used to cover the mouth of the test tube. (Be certain that the test tube mouth has an opening; the system should not be closed.)

5. Gradually heat the water in the beaker with a hot plate and watch for changes in temperature. As the liquid begins to boil, the temperature above the liquid will rise. When the temperature no longer rises but remains constant, record the temperature to the nearest 0.1°C (8). This is the observed boiling point. From the list in Table 4.2 identify your unknown by matching your observed boiling point with the compound whose boiling point corresponds the best (9).

6. Do as many boiling point determinations as required by your instructor.

7. Dispose of the liquid as directed by your instructor.

TABLE 4.2 BOILING POINTS OF SELECTED LIQUIDS

Liquid	Boiling Point (°C at 1 atm)
Acetone	56
Cyclohexane	81
Ethyl acetate	77
Hexane	69
Isopropyl alcohol	83
Methyl alcohol	65
1-Propanol	97

CHEMICALS AND EQUIPMENT

1. Aluminum foil
2. Boiling chips
3. Bunsen burner
4. Hot plate
5. Commercial melting point apparatus (if available)
6. Melting point capillary tubes
7. Rubber rings
8. Thiele tube melting point apparatus
9. Thermometer clamp
10. Glass tubing
11. Acetamide
12. Acetanilide
13. Acetone
14. Benzophenone
15. Benzoic acid
16. Biphenyl
17. Cyclohexane
18. Ethyl acetate
19. Hexane
20. Isopropyl alcohol
21. Lauric acid
22. Methyl alcohol
23. Naphthalene, pure
24. Naphthalene, impure
25. 1-Propanol
26. Stearic acid

EXPERIMENT 4

NAME _____ **SECTION** _____ **DATE** _____

PARTNER _____ **GRADE** _____

PRE-LAB QUESTIONS

1. For a given substance, are the condensation point of the gas and the boiling point of the liquid the same temperature? Explain.

2. Refer to Fig. 4.1. Estimate the boiling point of ethyl alcohol at 570 mm Hg.

3. Define sublimation. Give an example of a compound which undergoes sublimation.

4. How can water be brought to a boil at 60°C?

EXPERIMENT 4

NAME _____   SECTION _____   DATE _____

PARTNER _____   GRADE _____

<u>REPORT SHEET</u>

Melting point determination

	<u>Trial No. 1</u>	<u>Trial No. 2</u>
1. Unknown number	_____	_____
2. Temperature melting begins	_____ °C	_____ °C
3. Temperature melting ends	_____ °C	_____ °C
4. Melting point range	_____ °C	_____ °C
5. Identification of unknown	_____	_____

Purification of naphthalene by sublimation

6. Melting point range	_____ °C	_____ °C

Boiling point determination

7. Unknown number	_____	_____
8. Observed boiling point	_____ °C	_____ °C
9. Identification of unknown	_____	_____

POST-LAB QUESTIONS

1. In doing a melting point determination for benzoic acid, suppose a student took a melting point capillary tube that was previously used for a sample of lauric acid instead of a new one. How would this affect the melting point of the benzoic acid?

2. A student in Denver, Colorado, the "mile-high" city, carried out a boiling point determination for cyclohexane (b.p. 81°C) according to the procedure in this laboratory manual. Will this student's observed boiling point be the same as the value obtained by another student at sea level in New York City? Will it be lower or higher? Why?

3. Because proteins denature (change their folded structure) with heat, water cannot be removed from samples of eye lenses by heating to 100°C. How can the water be removed without destroying the integrity of the folded structure of the protein?

4. Cocaine is a white solid which melts at 98°C when pure. A forensic chemist working for the New York Police Department has a white solid believed to be cocaine. What can the chemist do to quickly determine whether the sample is cocaine and whether it is pure or a mixture?

Column and Paper Chromatography: Separation of Plant Pigments

EXPERIMENT 5

BACKGROUND

Chromatography is a widely used experimental technique by which a mixture of compounds can be separated into its individual components. Two kinds of chromatographic experiments will be explored. In column chromatography a mixture of components dissolved in a solvent is poured over a column of solid adsorbent and is eluted with the same or a different solvent. This is therefore a solid-liquid system, the stationary phase (the adsorbent) is solid and the mobile phase (the eluent) is liquid. In paper chromatography the paper adsorbs water from the atmosphere of the developing chromatogram. (The water is present in the air as vapor and it may be supplied as one component in the eluting solution). The water is the stationary phase. The (other) components of the eluting solvent is the mobile phase and carries with it the components of the mixture. This is a liquid-liquid system.

Column chromatography is used most conveniently for preparative purposes, when one deals with a relatively large amount of the mixture and the components need to be isolated in mg or g quantities. Paper chromatography on the other hand, is used mostly for analytical purposes. Microgram or even picogram quantities can be separated by this technique and they can be characterized by their R_f number. This number is an index of how far a certain spot moved on the paper.

$$R_f = \frac{\text{Distance of the center of the sample spot from the origin}}{\text{Distance of the solvent front from the origin}}$$

For example in Figure 5.1 the R_f values are as follows:

$$R_f \text{ (substance 1)} = 3.1 \text{ cm}/11.2 \text{ cm} = 0.28 \text{ and}$$
$$R_f \text{ (substance 2)} = 8.5 \text{ cm}/11.2 \text{ cm} = 0.76$$

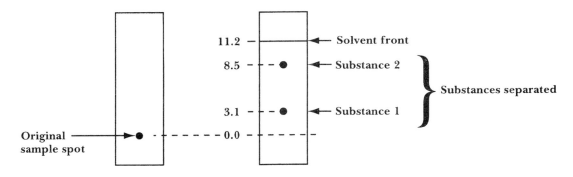

Figure 5.1 Illustration of chromatograms before and after elution.

Using the R_f values one is able to identify the components of the mixture with the individual components. The two main pigment components of tomato paste are ß-carotene (yellow-orange) and lycopene (red) pigments. Their structures are given below:

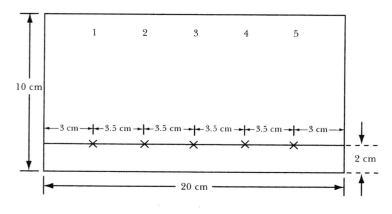

Lycopene

β-Carotene

The colors of these pigments are due to the numerous double bonds in their structure. When bromine is added to double bonds it saturates them and the color changes accordingly. In the tomato juice "rainbow" experiment we stir bromine water into the tomato juice. The slow stirring allows the bromine water to penetrate deeper and deeper into the cylinder in which the tomato juice was placed. As the bromine penetrates more and more double bonds will be saturated. Therefore, you may be able to observe a continuous change, a "rainbow" of colors starting with the reddish tomato color at the bottom of the cylinder where no reaction occurred since the bromine did not reach the bottom. Lighter colors will be observed on the top of the cylinder where most of the double bonds have been saturated.

OBJECTIVE

1. To compare separation of components of a mixture by two different techniques.
2. To demonstrate the effect of bromination on plant pigments of tomato juice.

PROCEDURE

Part A. Paper chromatography

1. Obtain a sheet of Whatman no.1 filter paper, cut to size.

Figure 5.2 Preparation of chromatographic paper for spotting.

2. Plan the spotting of the samples as illustrated on Fig. 5.2. Five spots will be applied. The first and fifth will be ß-carotene solutions supplied by your instructor. The second, third and fourth lanes will have your tomato paste extracts in different concentrations. Use a pencil to mark lightly the spots according to Fig. 5.2.

3. Pigments of tomato paste will be **extracted** in two steps.
   a. Weigh about 10 g of tomato paste in a 50-mL beaker. Add 15 mL of 95% ethanol. Stir the mixture vigorously with a spatula until the paste will not stick to the stirrer. Place a small amount of glass wool (the size of a pea) in a small funnel blocking the funnel exit. Place the funnel into a 50-mL Erlenmeyer flask and pour the tomato paste–ethanol mixture into the funnel. When the filtration is completed squeeze the glass wool lightly with your spatula. In this step we removed the water from the tomato paste and the aqueous components are in the filtrate, which we discard. The residue in the glass wool will be used to extract the pigments.
   b. Place the residue in the glass wool in a 50-mL beaker. Add 10 mL petroleum ether and stir the mixture for about two minutes to extract the pigments. Filter the extract as before through a new funnel with glass wool blocking the exit, into a new and clean 50-mL beaker. Place the beaker under the hood on a hot plate (or water bath) and evaporate it to about 1 mL volume. Use low heat and take care not to evaporate all the solvent. After evaporation cover the beaker with aluminum foil.

Figure 5.3 Withdrawing samples by a capillary tube.

4. **Spotting.** Place your chromatographic paper on a clean area (another filter paper) in order not to contaminate it. Use separate capillaries for your tomato paste extract and for the ß-carotene solution. First apply your capillary to the extracted pigment by dipping it into the solution as illustrated in Fig. 5.3. Apply the capillary lightly to the chromatographic paper by touching sequentially the spots marked 2, 3 and 4. Make sure you apply only small spots, not larger than 2 mm diameter, by quickly withdrawing the capillary from the paper each time you touch it. (See Fig. 5.4).

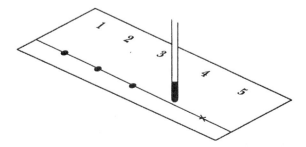

Figure 5.4 Spotting.

While allowing the spots to dry, use your second capillary to apply spots of ß-carotene in lanes 1 and 5. Return to the first capillary and apply another spot of the extract on top of the spots of lanes 3 and 4. Let them dry (Fig. 5.5). Finally apply one more spot on top of lane 4. Let the spots dry. The unused extract in your beaker should be covered with aluminum foil. Place it in your drawer in the dark to save it for the second part of your experiment.

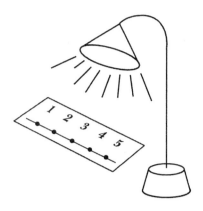

Figure 5.5 Drying chromatographic spots.

5. **Developing the Paper Chromatogram**. Curve the paper into a cylinder and staple the edges above the 2 cm line as it is shown in Fig. 5.6.

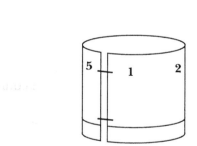   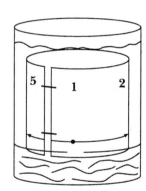

Figure 5.6 Stapling.   **Figure 5.7 Developing the chromatogram.**

6. Pour 20 mL of the eluting solvent (petroleum ether : toluene : acetone in 45:1:5 ratio, supplied by your instructor) into a 600-mL beaker.

7. Place the stapled chromatogram into the 600-mL beaker the spots being at the bottom near the solvent surface but **not covered by it**. Cover the beaker with aluminum foil. Allow the solvent front to migrate up to 0.5 - 1 cm below the edge of the paper. This may take from 15 min. to one hr. Make certain by frequent inspection that the **solvent front does not run over the edge of the paper**. Remove the chromatogram from the beaker when the solvent front reaches 0.5 - 1 cm from the edge. Proceed to step 11.

Part B.

8. **Column Chromatography**. While you are waiting for the chromatogram to develop (step 7) you can perform the column chromatography experiment. Take a 25-mL buret. (You may use a chromatograhic column, if available, of 1.6 cm diameter and about 13 cm long, see Fig. 5.8. If you use the column all subsequent quantities below should be doubled).

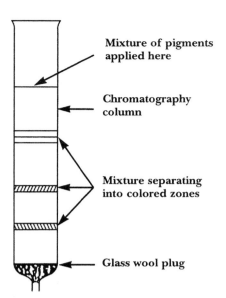

Mixture of pigments applied here

Chromatography column

Figure 5.8 Chromatographic column.

Mixture separating into colored zones

Glass wool plug

Add a small piece of glass wool and with the aid of a glass rod push it down near the stopcock. Add 15-16 mL of petroleum ether to the buret. Open the stopcock slowly and allow the solvent to fill the tip of the buret. Close the stopcock. You should have 12-13 mL of solvent above the glass wool. Weigh 20 g of aluminum oxide (alumina) in a 100-mL beaker. Place a small funnel on top of your buret. Pour the alumina into the buret. Allow the alumina to settle to form a 20 cm column. Drain the solvent but *do not allow the column to run dry. Always have at least a 0.5 mL of clear solvent on top of the column.* If alumina adheres to the walls of the buret wash it down with more solvent.

9. Transfer by pipet 0.5-1 mL of the extract you stored in your drawer onto the column. The pipet containing the extract should be placed near the surface of the solvent on top of the column. Touching the walls of the buret with the tip of the pipet allow the extract to drain slowly on top of the column. Open the stopcock slightly. Allow the sample to enter the column, *but make sure there is a small amount of solvent on top of the column. (The column should never run dry).* Add 10 or more mL petroleum ether and wash the sample into the column by open-ing the stopcock and collecting the eluted solvent in a beaker.

10. As the solvent elutes the sample, you observe the migration of the pigment and their separa-tion into at least two bands. When the fastest moving pigment band reaches near the bottom of the column, close the stopcock and observe the color of the pigment bands and how far they migrated from the top of the column. This concludes the column chromatographic part of the experiment. Discard your solvent in a bottle supplied by your instructor for a later re-distillation.

11. Meanwhile your paper chromatogram has developed. You must remove the filter paper from the 600-mL beaker before the solvent front reaches the edges of the paper. Mark the position of the solvent front with a pencil. Put the paper standing on its edges under the hood and let it dry.

Part C.

12. **Tomato Juice "Rainbow".** While waiting for the paper to dry you can perform the following short experiment. Weigh about 15 g tomato paste in a beaker. Add about 30 mL of water and stir. Transfer the tomato juice into a 50-mL cylinder and with the aid of a pipet add 5 mL of saturated bromine water dropwise. With a glass rod stir very gently the solution. Observe the colors and their positions in the cylinder.

Part A cont.

13. Remove the staples from the dried chromatogram. Mark the spots of the pigments by circling them with a pencil. Note the colors of the spots. Measure the distance of the center of each spot from its origin. Calculate the R_f values.

14. If the spots on the chromatogram are faded we can visualize them by exposing the chromatogram to iodine vapor. Place your chromatogram into a wide mouth jar containing a few iodine crystals. Cap the jar and warm it slightly on a hot plate to enhance the sublimation of iodine. The iodine vapor will interact with the faded pigment spots and make them visible. After a few minutes exposure to iodine vapor remove the chromatogram and mark the spots **immediately** with pencil. The spots will fade again with exposure to air. Measure the distance of the center of the spots from the origin and calculate the R_f values.

CHEMICALS AND EQUIPMENT

1. Melting point capillaries open at both ends
2. 25-mL buret or chromatographic column
3. Glass wool
4. Whatman no.1 filter papers, 10 × 20 cm, cut to size
5. Heat lamp (optional)
6. Stapler
7. Hot plate with or without water bath
8. Tomato paste
9. Aluminum oxide (alumina)
10. Petroleum ether (b.p. 30-60°C)
11. 95% ethanol
12. Toluene
13. Acetone
14. 0.5% ß-carotene in petroleum ether
15. Saturated bromine water.
16. Iodine crystals
17. Ruler

EXPERIMENT 5

NAME _____   SECTION _____   DATE _____

PARTNER _____   GRADE _____

PRE-LAB QUESTIONS

1. (a) What is the "stationary phase" in column chromatography?
    (b) What is the "mobile phase" in column chromatography?

2. The structures of the two main pigments, lycopene and ß-carotene are given in the first part (Background):

    (a) Are these pigments hydrocarbons?

    (b) What functional groups are present in these pigments?

    (c) What solvents will be good for these pigments: polar or non-polar?

3. Write the structure of ß-carotene after it completely reacted with Br_2.

EXPERIMENT 5

NAME _____ SECTION _____ DATE _____

PARTNER _____ GRADE _____

REPORT SHEET

Paper chromatography

Sample	Distance from origin to solvent front (cm) (a)	Distance from origin to center of spot (cm) (b)	R_f (b)/(a)	Color

ß-carotene

lane 1

lane 5

Tomato extract

lane 2 (a)

       (b)

       (c)

       (d)

lane 3 (a)

       (b)

       (c)

       (d)

lane 4 (a)

       (b)

       (c)

       (d)

Column chromatography

Number of bands	Distance migrated from top of the column (cm)	Color
1		
2		
3		

Describe the colors observed in the tomato juice "rainbow" experiment, starting from the bottom of the cylinder:

1. red

2.

3.

4.

5.

6.

POST-LAB QUESTIONS

1. Did your tomato paste contain lycopene? What support is there for your answer?

2. Which chromatographic technique gave you a better separation? Explain.

3. What is the effect of the amount of sample applied to the paper on the separation of the tomato pigments? Compare the results on lanes 2, 3 and 4 of the paper chromatogram.

4. Based on the R_f value you calculated for lycopene, how far would this pigment travel if the solvent front moved 20 cm?

5. Tomato juice is red. What does this tell you about its pigment composition?

Isolation of Caffeine from Tea Leaves

BACKGROUND

Many organic compounds are obtained from natural sources through extraction. This method takes advantage of the solubility characteristics of a particular organic substance with a given solvent. In the experiment here, caffeine is readily soluble in hot water and is thus, separated from the tea leaves. Caffeine is one of the main substances that make up the water solution called tea. Besides being found in tea leaves, caffeine is present in coffee, kola nuts and cocoa beans. As much as 5% by weight of the leaf material in tea plants consists of caffeine.

The caffeine structure is shown below. It is classified as an

alkaloid, meaning that with the nitrogen present, the molecule has base characteristics (alkali-like). In addition the molecule has the purine ring system, a framework which plays an important role in living systems.

Caffeine is the most widely used of all the stimulants. Small doses of this chemical (50 to 200 mg) can increase alertness and reduce drowsiness and fatigue. The popular "No-Doz" tablet has caffeine as the main ingredient. In addition it affects blood circulation since the heart is stimulated and blood vessels are relaxed (vasodilation). It also acts as a diuretic. There are side effects. Large doses of over 200 mg can result in insomnia, restlessness, headaches and muscle tremors ("coffee nerves"). Continued, heavy use may bring on physical dependence. (How many of you know someone who cannot function in the morning until they have that first cup of coffee?)

Tea leaves consist primarily of cellulose since this is the principal structural material of all plant cells. Fortunately, the cellulose is insoluble in water, so that by using a hot water extraction, the more soluble caffeine can be separated. Also dissolved in water are complex substances called tannins. These are colored phenolic compounds of high molecular weight (500 to 3000) that have acidic behavior. If a basic salt such as Na_2CO_3 is added to the water solution, the tannins can react to form a salt. These salts are insoluble in organic solvents such as chloroform or methylene chloride, but are soluble in water.

Although caffeine is soluble in water (2 g/100 g of cold water), it is more soluble in the organic solvent methylene chloride (14 g/100 g). Thus, caffeine can be extracted from the basic tea solution with methylene chloride, but the sodium salts of the tannins remain behind in the aqueous solution. Evaporation of the methylene chloride yields crude caffeine; the crude material can be purified by sublimation (see Experiment 4).

OBJECTIVES

1. To demonstrate the isolation of a natural product.
2. To learn the techniques of extraction.
3. To use sublimation as a purification technique.

PROCEDURE

The isolation of caffeine from tea leaves follows the scheme below:

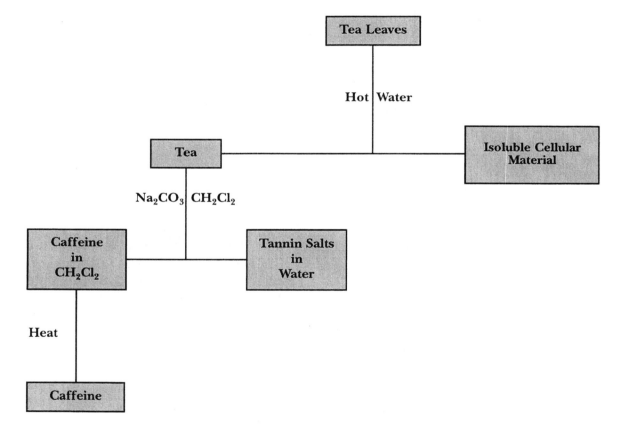

1. Carefully open two commercial tea bags (try not to tear the paper) and weigh the contents to the nearest 0.001 g. Record this weight (1). Place the tea leaves back into the bags, close and secure the bags with staples.

2. Into a 150-mL beaker place the tea bags so that they lie flat on the bottom. Add 30 mL of distilled water and 2.0 g of anhydrous Na_2CO_3; heat the contents with a hot plate, keeping a gentle boil, for 20 min. While the mixture is boiling, keep a watch glass on the beaker. Hold the tea bags under water by occasionally pushing them down with a glass rod.

3. Decant the hot liquid into a 50-mL Erlenmeyer flask. Wash the tea bags with 10 mL of hot water, carefully pressing the tea bag with a glass rod; add this wash water to the tea extract. (If

any solids are present in the tea extract, filter them by gravity to remove.) Cool the combined tea extract to room temperature. The tea bags may be discarded.

4. Transfer the cool tea extract to a 125-mL separatory funnel supported on a ring stand with a ring clamp.

5. Carefully add 5.0 mL of methylene chloride to the separatory funnel. Stopper the funnel and lift the funnel from the ring clamp; hold the funnel with two hands as shown in Fig. 6.1. By holding the stopper in place with one hand, invert the funnel. *Make certain the stopper is held tightly and no liquid is spilled; open the stopcock, being sure to point the opening away from you and your neighbors.* Built-up pressure caused by gases accumulating inside will be released. Now close the stopcock and gently mix the contents by inverting two or three times. Again release any pressure by opening the stopcock as before.

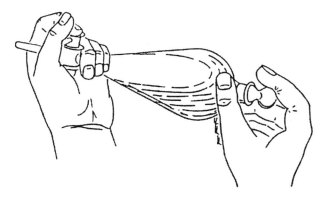

Figure 6.1 Using the separatory funnel.

6. Return the separatory funnel to the ring clamp, remove the stopper and allow the aqueous layer to separate from the methylene chloride layer (Fig. 6.2). You should see two distinct layers form after a few minutes with the methylene chloride layer at the bottom. Sometimes an emulsion may form at the juncture of the two layers. The emulsion often can be broken by gently swirling the contents or by gently stirring the emulsion with a glass rod.

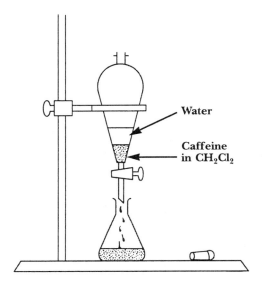

Water

Caffeine in CH_2Cl_2

Figure 6.2 Separation of the aqueous layer and the methylene chloride layer in the separatory funnel.

7. Carefully drain the lower layer into a 25-mL Erlenmeyer flask. Try not to include any water with the methylene chloride layer; careful manipulation of the stopcock will prevent this.

8. Repeat the extraction with an additional 5.0 mL of methylene chloride. Combine the separated bottom layer with the methylene chloride layer obtained in step 7.

9. Add 0.5 g of anhydrous Na_2SO_4 to the combined methylene chloride extracts. Swirl the flask. The anhydrous salt is a drying agent and will remove any water that may still be present.

10. Weigh a 25-mL side-arm filter flask containing one or two boiling stones. Record this weight (2). By means of a gravity filtration, filter the methylene chloride-salt mixture into the pre-weighed flask. Rinse the salt on the filter paper with an additional 2.0 mL of methylene chloride.

11. Remove the methylene chloride by evaporation *in the hood.* Be careful not to overheat the solvent since it may foam over. The solid residue which remains after the solvent is gone is the crude caffeine. Reweigh the cooled flask (3). Calculate the weight of the crude caffeine by subtraction (4) and determine the percent yield (5).

12. Take a melting point of your solid. First scrape the caffeine from the bottom and sides of the flask with a microspatula and collect a sample of the solid in a capillary tube (review Experiment 4 for the technique). Pure caffeine melts at 238°C. Compare your melting point (6) to the literature value.

13. At the option of your instructor, the caffeine may be purified further. The caffeine may be sublimed directly from the flask with a cold finger condenser (Fig. 6.3). Carefully insert the cold finger condenser into a no. 2 neoprene adapter (use a drop of glycerine as a lubricant). Adjust the tip of the cold finger to 1 cm from the bottom of the flask. Clean any glycerine remaining on the cold finger with a Kimwipe and acetone; the cold finger surface must be clean and dry. Connect the cold finger to a faucet by latex tubing (water in the upper tube; water out the lower tube). Connect the side-arm filter flask to a water aspirator with vacuum tubing, installing a trap between the aspirator and the sublimation set-up (Fig. 6.3). When you turn the water on, press the cold finger into the filter flask until a good seal is made. Gently heat the bottom of the filter flask which holds the caffeine with a microburner (hold the base of the microburner); move the flame back and forth and along the sides of the flask. **Do not allow the sample to melt;** if the sample melts, stop heating and allow to cool before continuing. When the sublimation is complete, disconnect the heat and allow the system to cool; leave the aspirator connected and the water running.

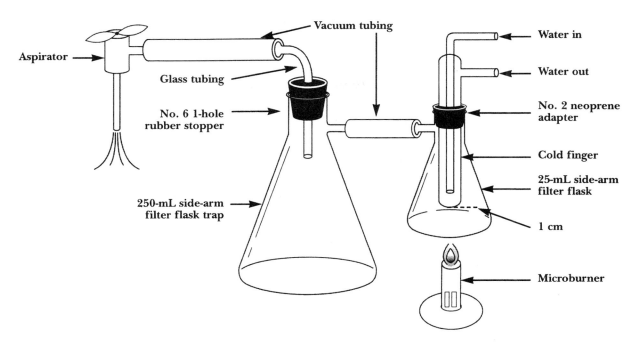

Figure 6.3 Sublimation apparatus connected to an aspirator.

14. When the system has reached room temperature, carefully disconnect the aspirator from the side-arm filter flask by removing the vacuum tubing from the side-arm. Turn off the water to the cold finger. Carefully remove the cold finger from the flask along with the neoprene adapter without dislodging any crystals. Scrape the sublimed caffeine onto a pre-weighed piece of weighing paper (7). Reweigh (8); determine the weight of caffeine (9). Calculate the percent recovery (10). Determine the melting point (11).

15. Collect the caffeine in a sample vial and submit it to your instructor.

CHEMICALS AND EQUIPMENT

1. Boiling chips
2. Cold finger condenser
3. Filter paper, (7.0) fast flow, Whatman no. 1
4. Hot plate
5. 125-mL separatory funnel with stopper
6. Melting point capillaries
7. No. 2 neoprene adapter
8. 25-mL side-arm filter flask
9. Small sample vials
10. Tea bags
11. Tubing: latex, 2 ft.; vacuum, 2 ft.
12. 250-mL trap: 250-mL side-arm filter flask fitted with a no. 6 one-hole rubber stopper containing a piece of glass tubing (10 cm long × 7 mm OD)
13. Anhydrous sodium sulfate, Na_2SO_4
14. Anhydrous sodium carbonate, Na_2CO_3
15. Methylene chloride, CH_2Cl_2
16. Stapler

EXPERIMENT 6

NAME _____ SECTION _____ DATE _____

PARTNER _____ GRADE _____

PRE-LAB QUESTIONS

1. Name some sources of caffeine.

2. What part of the caffeine structure gives the alkaloid properties?

3. How does caffeine affect an individual?

4. How can you test the purity of a sample of caffeine?

EXPERIMENT 6

NAME _____ SECTION _____ DATE _____

PARTNER _____ GRADE _____

REPORT SHEET

1. Weight of tea in 2 tea bags   _____ g

2. Weight of 25-mL side-arm filter flask and boiling stones   _____ g

3. Weight of flask, boiling stones and crude caffeine   _____ g

4. Weight of caffeine: $(3) - (2)$ _____ g

5. Percent yield: $[(4)/(1)] \times 100$ _____ %

6. Melting point of your crude caffeine   _____ °C

7. Weight of weighing paper   _____ g

8. Weight of sublimed caffeine and paper   _____ g

9. Weight of caffeine: $(8) - (7)$ _____ g

10. Percent recovery: $[(9)/(4)] \times 100$ _____ %

11. Melting point of sublimed caffeine   _____ °C

POST-LAB QUESTIONS

1. Compare the melting points of the crude and sublimed samples of caffeine. Account for the differences between the melting points. How do they compare to the literature value?

2. What other compounds were extracted along with caffeine in the hot water extract (tea)? How were these compounds separated from the caffeine?

3. In the procedure, why was the caffeine eventually recovered from the methylene chloride extract and not directly from the water?

4. How do you know that methylene chloride has a greater density than water?

Identification of Hydrocarbons

BACKGROUND

The number of known organic compounds total into the millions. Of these compounds the simplest types are those which contain only hydrogen and carbon atoms. These are known as hydrocarbons. Because of the number and variety of hydrocarbons that can exist, some means of classification is necessary.

One means of classification depends on the way in which carbon atoms are connected. Aliphatic hydrocarbons are compounds consisting of carbons linked either in a single chain or in a branched chain. Cyclic hydrocarbons are compounds having the carbon atoms linked in a closed polygon. For example, hexane and 2-methylpentane are aliphatic molecules, while cyclohexane is a cyclic system.

$$CH_3CH_2CH_2CH_2CH_2CH_3$$

Hexane

$$CH_3CHCH_2CH_2CH_3$$
$$|$$
$$CH_3$$

2-Methylpentane

Cyclohexane

Another means of classification depends on the type of bonding that exists between carbons. Hydrocarbons which contain only carbon to carbon single bonds are called alkanes. These are also referred to as saturated molecules. Hydrocarbons containing at least one carbon to carbon double bond are called alkenes, and those compounds with at least one carbon to carbon triple bond are alkynes. These are compounds that are referred to as unsaturated molecules. Finally, a class of cyclic hydrocarbons which contain a closed loop (sextet) of electrons are called aromatic (see your text for further details). Table 7.1 distinguishes between the families of hydrocarbons.

With so many compounds possible, identification of the bond type is an important step in establishing the molecular structure. Quick, simple tests on small samples can establish the physical and chemical properties of the compounds by class.

Some of the observed physical properties of hydrocarbons result from the non-polar character of the compounds. In general, hydrocarbons do not mix with polar solvents such as water or ethyl alcohol. On the other hand, hydrocarbons mix with relatively non-polar solvents such as ligroin (a mixture of alkanes), carbon tetrachloride or dichloromethane. Since the density of most hydrocarbons is less than that of water, they will float. Crude oil and crude oil products (home heating oil and gasoline) are mixtures of hydrocarbons; these substances, when spilled on water, spread quickly along the surface because they are insoluble in water.

TABLE 7.1 TYPES OF HYDROCARBONS

Class		Characteristic Bond Type	Example	
I. Aliphatic				
1. Alkane[a]	$-\overset{\mid}{\underset{\mid}{C}}-\overset{\mid}{\underset{\mid}{C}}-$	single	$CH_3CH_2CH_2CH_2CH_2CH_3$	hexane
2. Alkene[b]	$\diagdown C = C \diagup$	double	$CH_3CH_2CH_2CH_2CH=CH_2$	1-hexene
3. Alkyne[b]	$-C\equiv C-$	triple	$CH_3CH_2CH_2CH_2C\equiv CH$	1-hexyne
II. Cyclic				
1. Cycloalkane[a]	$-\overset{\mid}{\underset{\mid}{C}}-\overset{\mid}{\underset{\mid}{C}}-$	single		cyclohexane
2. Cycloalkene[b]	$\diagdown C = C \diagup$	double		cyclohexene
3. Aromatic				benzene
			CH_3	toluene

[a] Saturated   [b] Unsaturated

The chemical reactivity of hydrocarbons is determined by the type of bond in the compound. Although saturated hydrocarbons, alkanes, will burn (undergo combustion), they are generally unreactive to most reagents. (Alkanes do undergo a substitution reaction with halogens, but require ultraviolet light.) Unsaturated hydrocarbons, alkenes and alkynes, not only burn, but also react by addition of reagents to the double or triple bonds. The addition products become saturated, with fragments of the reagent becoming attached to the carbons of the multiple bond. Aromatic compounds, with a higher carbon to hydrogen ratio than non-aromatic compounds, burn with a sooty flame as a result of unburned carbon particles being present. These compounds undergo substitution in the presence of catalysts rather than an addition reaction.

1. **Combustion.** The major component in "natural gas" is the hydrocarbon methane. Other hydrocarbons used for heating or cooking purposes are propane and butane. The products from combustion are carbon dioxide and water (heat is evolved, also).

$$CH_4 + 2O_2 \rightarrow CO_2 + 2H_2O$$

$$(CH_3)_2CHCH_2CH_3 + 8O_2 \rightarrow 5CO_2 + 6H_2O$$

2. **Reaction with bromine.** Unsaturated hydrocarbons react rapidly with bromine in a solution of carbon tetrachloride or cyclohexane. The reaction is the addition of the elements of bromine to the carbons of the multiple bonds.

$$CH_3CH{=}CHCH_3 + Br_2 \rightarrow \underset{\text{Colorless}}{CH_3CH{-}CHCH_3}$$

$$\overset{\text{Br}\quad\text{Br}}{\underset{|\quad\;\;|}{}}$$

Red

$$CH_3C{\equiv}CCH_3 + 2\,Br_2 \rightarrow CH_3C{-}CCH_3$$

Red   Colorless

The bromine solution is red; the product which has the bromine atoms attached to carbon is colorless. Thus, a reaction has taken place when there is a loss of color from the bromine solution and a colorless solution remains. Since alkanes have only single $C{-}C$ bonds present, no reaction with bromine is observed; the red color of the reagent would persist when added. Aromatic compounds resist addition reactions because of their "aromaticity": the possession of a closed loop (sextet) of electrons. These compounds react with bromine in the presence of a catalyst such as iron filings or aluminum chloride.

Note that a substitution reaction has taken place and the gas HBr is produced.

3. **Reaction with concentrated sulfuric acid.** Alkenes react with cold concentrated sulfuric acid by addition. Alkyl sulfonic acids form as products and are soluble in H_2SO_4.

$$CH_3{-}CH{=}CH{-}CH_3 + HOSO_2OH \rightarrow CH_3{-}CH{-}CH{-}CH_3$$
$$(H_2SO_4) \qquad\qquad\quad \underset{H\quad\;\;OSO_2OH}{|\quad\;\;|}$$

Saturated hydrocarbons are unreactive (additions are not possible); alkynes react slowly and require a catalyst ($HgSO_4$); aromatic compounds also are unreactive since addition reactions are difficult.

4. **Reaction with potassium permanganate.** Dilute neutral or alkaline solutions of $KMnO_4$ oxidize unsaturated compounds. Alkanes and aromatic compounds are generally unreactive. Evidence that a reaction has occurred is by the loss of the purple color of $KMnO_4$ and the formation of the brown precipitate manganese dioxide, MnO_2.

$$3CH_3-CH=CH-CH_3 + 2KMnO_4 + 4H_2O \rightarrow 3CH_3-CH-CH-CH_3 + 2MnO_2 + 2KOH$$

Purple   OH   OH   Brown

Note that the product formed from an alkene is a glycol or diol.

OBJECTIVES

1. To investigate the physical properties, solubility and density, of some hydrocarbons.
2. To compare the chemical reactivity of an alkane, an alkene and an aromatic compound.
3. To use physical and chemical properties to identify an unknown.

PROCEDURE

Assume the organic compounds are highly flammable. Use only small quantities. Keep away from open flames. Assume the organic compounds are toxic and can be absorbed through the skin. Avoid contact; wash if any chemical spills on your person. Handle concentrated sulfuric acid carefully. Flush with water if any spills on your person. Potassium permanganate and bromine are toxic; bromine solutions are also corrosive. Although the solutions are dilute, they may cause burns to the skin. Wear gloves when working with these chemicals.

General instructions

1. The hydrocarbons hexane, cyclohexene and toluene (alkane, alkene and aromatic) are available in dropper bottles.

2. The reagents 1% Br_2 in cyclohexane, 1% aqueous $KMnO_4$ and concentrated H_2SO_4 are available in dropper bottles.

3. Unknowns are in dropper bottles labeled A, B and C. They may include an alkane, an alkene or an aromatic compound.

4. Record all data and observations in the appropriate places on the Report Sheet.

5. Dispose of all organic wastes as directed by the instructor. *Do not pour into the sink!*

Physical properties of hydrocarbons

1. A test tube of 100 × 13 mm will be suitable for this test. When mixing the components, grip the test tube between thumb and forefinger; it should be held firmly enough to keep from slipping but loosely enough so that when the third and fourth fingers tap it, the contents will be agitated enough to mix.

2. **Water solubility of hydrocarbons.** Label six test tubes with the name of the substance to be tested. Place into each test tube 5 drops of the appropriate hydrocarbon: hexane, cyclo-hexene, toluene, unknown A, unknown B, unknown C. Add about 5 drops of water dropwise

into each test tube. Is there any separation of components? Which component is on the bottom; which component is on the top? Mix the contents as described above. What happens when the contents are allowed to settle? What do you conclude about the density of the hydrocarbon? Record your observations. Save these solutions for comparison with the next part.

3. **Solubility of hydrocarbons in ligroin.** Label six test tubes with the name of the substance to be tested. Place into each test tube 5 drops of the appropriate hydrocarbon: hexane, cyclohexene, toluene, unknown A, unknown B, unknown C. Add about 5 drops of ligroin dropwise into each test tube. Is there a separation of components? Is there a bottom layer and top layer? Mix the contents as described above. Is there any change in the appearance of the contents before and after mixing? Compare these test tubes to those from the previous part. Can you make any conclusion about the density of the hydrocarbon? Record your observations.

Chemical properties of hydrocarbons

1. **Combustion.** The instructor will demonstrate this test in the fume hood. Place 5 drops of each hydrocarbon and unknown on separate watch glasses. Carefully ignite each sample with a match. Observe the flame and color of the smoke for each of the samples. Record your observations on the Report Sheet.

2. **Reaction with bromine.** Label six clean, dry test tubes with the name of the substance to be tested. Place into each test tube 5 drops of the appropriate hydrocarbon: hexane, cyclohexene, toluene, unknown A, unknown B, unknown C. Carefully add, dropwise and with shaking, 1% Br_2 in cyclohexane. (**CAUTION!** *Use in hood; wear gloves when using this chemical.*) Keep count of the number of drops needed to have the color persist; do not add more than 10 drops. Record your observations.

    To the test tube containing toluene, add 5 more drops of 1% Br_2 solution and a small quantity of iron filings; shake the mixture. Place a piece of moistened blue litmus paper on the test tube opening. Record any change in the color of the solution and the litmus paper.

3. **Reaction with $KMnO_4$.** Label six clean, dry test tubes with the name of the substance to be tested. Place into each test tube 5 drops of the appropriate hydrocarbon: hexane, cyclohexene, toluene, unknown A, unknown B, unknown C. Carefully add, dropwise and with shaking 1% aqueous $KMnO_4$ solution. Keep count of the number of drops needed to have the color persist; do not add more than 10 drops. Record your observations.

4. **Reaction with concentrated H_2SO_4.** Label six clean, dry test tubes with the name of the substance to be tested. Place into each test tube 5 drops of the appropriate hydrocarbon: hexane, cyclohexene, toluene, unknown A, unknown B, unknown C. Place all of the test tubes in an ice bath. Wear gloves and carefully add with shaking, 3 drops of cold, concentrated sulfuric acid to each test tube. Note whether heat is evolved by feeling the test tube. Note whether the solution has become homogeneous or whether a color is produced. (The evolution of heat or the formation of a homogeneous solution or the appearance of a color is evidence that a reaction has occurred.) Record your observations.

5. **Unknowns.** By comparing the observations you make for your unknowns with that of the known hydrocarbons, you can identify unknowns A, B and C. Record their identities on your Report Sheet.

CHEMICALS AND EQUIPMENT

1. 1% aqueous $KMnO_4$
2. 1% Br_2 in cyclohexane
3. Blue litmus paper
4. Concentrated H_2SO_4
5. Cyclohexene
6. Hexane
7. Iron filings or powder
8. Test tubes
9. Ligroin
10. Toluene
11. Unknowns A, B, and C
12. Watch glasses
13. Ice

EXPERIMENT 7

NAME _____ SECTION _____ DATE _____

PARTNER _____ GRADE _____

PRE-LAB QUESTIONS

1. Why is hexane known as a hydrocarbon?

2. Show the structural feature that distinguishes whether a hydrocarbon is an

   alkane

   alkene

   alkyne

   aromatic

3. Hydrocarbons do not mix with water. What accounts for this property?

EXPERIMENT 7

NAME _____ SECTION _____ DATE _____

PARTNER _____ GRADE _____

<u>REPORT SHEET</u>

Physical properties of hydrocarbons

Hydrocarbon	H_2O		Ligroin	
	Solubility	Density	Solubility	Density
Hexane				
Cyclohexene				
Toluene				
Unknown A				
Unknown B				
Unknown C				

Chemical properties of hydrocarbons

Hydrocarbon	Combustion	Bromine	KMnO$_4$ Test	H$_2$SO$_4$ Test
Hexane				
Cyclohexene				
Toluene				
Unknown A				
Unknown B				
Unknown C				

Unknown A is _____.

Unknown B is _____.

Unknown C is _____.

POST-LAB QUESTIONS

1. Write the structure of the major organic product for the following reactions:

   a. $CH_3-CH=CH_2 + Br_2 \rightarrow$

   b. 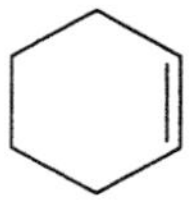 $+ Br_2 \rightarrow$

   c. $CH_3-CH=CH-CH_3 + KMnO_4 + H_2O \rightarrow$

   d. $CH_3CH_2-CH=CH-CH_2CH_3 + H_2SO_4 \rightarrow$

2. Of the compounds used in this experiment (hexane, cyclohexene, toluene) which was the most reactive with bromine solution? Explain your conclusion.

3. On the basis of the chemical tests used in this experiment, is it possible to *clearly* distinguish between an alkane and an alkene? Explain your conclusion.

Dehydration of 2-Methylcyclohexanol: an Elimination Reaction

BACKGROUND

The dehydration of alcohols can be used to prepare alkenes. The elimination of the elements of water from the alcohol is acid-catalyzed and uses strong, concentrated mineral acids, such as phosphoric or sulfuric acids.

$$-\underset{\underset{H}{|}}{C}-\underset{\underset{OH}{|}}{C}- \ + \ H^+ \ \rightarrow \ -\underset{|}{C}=\underset{|}{C}- \ + \ H_3O^+$$

Alcohol   Alkene

In this experiment 2-methylcyclohexanol undergoes dehydration in the presence of phosphoric acid to yield a mixture of alkenes: 1-methylcyclohexene and 3-methylcyclohexene.

2-Methylcyclohexanol   1-Methylcyclohexene   3-Methylcyclohexene

The relative percentages of the alkenes in the mixture can be determined by gas chromatography. Using Zaitsev's rule, the expectation is that 1-methylcyclohexene, the alkene with the more substituted double bond, should predominate.

vs.   miNoR

major

3 substituents   2 substituents

The acid serves to protonate the hydroxyl group; this allows neutral water to dissociate, a better leaving group than the hydroxyl anion. Regiochemistry is determined through proton loss: loss of H_a gives 1-methylcyclohexene; loss of H_b gives 3-methylcyclohexene.

Under the experimental conditions, the equilibrium is shifted to products by distillation of the alkenes as they are formed. This also minimizes the loss of product by acid catalyzed polymerization. As the distillation progresses, however, water and phosphoric acid co-distill with the alkenes (b. p. 103 - 111°C). Also, if the distillation is not done carefully, some 2-methylcyclohexanol (b. p. 165°C) may contaminate the product. Washing with saturated sodium chloride removes the phosphoric acid, and drying with anhydrous sodium sulfate takes out the water. A final distillation gives the product free of contaminants.

Simple test tube reactions can be used to give qualitative tests for the presence of double bonds in organic compounds (see Experiment 7). Bromine solutions, red in color, are decolorized by alkenes. Aqueous potassium permanganate, a purple solution, loses its color with alkenes and gives the brown precipitate of manganese dioxide.

OBJECTIVES

1. To demonstrate the dehydration of an alcohol.
2. To use dehydration to prepare alkenes.
3. To use gas chromatography as an analytical technique to determine product composition and to verify Zaitsev's rule.
4. To show qualitative tests for unsaturation.

PROCEDURE

Reaction

1. Weigh a 50-mL round bottom flask to the nearest 0.01 g; record the weight on the Report Sheet (1). Add 15.0 mL of 2-methylcyclohexanol (specific gravity = 0.933; MW = 114.2). Reweigh the flask (2) and by subtraction, determine the mass of 2-methylcyclohexanol (3); record these dehts to the nearest 0.01 g on the Report Sheet.

2. Carefully add 6.0 mL of 85% phosphoric acid. Swirl the flask to thoroughly mix the liquids; add two (2) boiling stones.

CAUTION! *Concentrated phosphoric acid can cause severe burns if you get any on your skin or clothing. Wear gloves and dispense in the hood. If you come into contact with any acid, flush the area immediately with a large amount of water. Inform your instructor and seek medical help.*

3. Assemble an apparatus for distillation (see Fig. 3.2, pg. 24) using a water-cooled condenser and a 25-mL round bottom receiving flask.

Dehydration

1. Carefully turn the water spigot on and allow water to circulate through the condenser. A gentle stream of water is sufficient.

2. Begin heating the mixture using a heating mantle, controlled by a Variac. Monitor the temperature of the distilling vapor; allow to go no higher than 111°C, the boiling point of 1-methylcyclohexene.

3. Continue the distillation until material no longer distills. There should be only a few milliliters of residue remaining in the distilling flask. Should white fumes appear at this point, stop heating and remove the heating mantle (the acids are starting to decompose). Note that the material in the receiving flask shows two (2) layers.

CAUTION! *Allow the distilling flask to cool before disassembly. The residue should be neutralized by cautiously adding cold, 10% sodium carbonate solution to the flask. The material will froth! When the solution is basic to litmus, flush down the sink with plenty of water.*

Isolation

1. Transfer the contents of the receiving flask to a 125-mL separatory funnel (see Figs. 6.1 and 6.2, pg. 59, for techniques). Slowly add 5.0 mL of 10% sodium carbonate solution. The acid present will generate gas as it is neutralized. Vent by opening the stopcock to release the pressure as you gently swirl or shake the mixture.

2. Place the separatory funnel on a ring clamp (Fig. 6.2, pg. 59) and allow the layers to separate.

3. Remove the stopper, open the stopcock and draw off the lower aqueous layer. Pour the remaining upper organic layer out the top of the separatory funnel into a clean, dry 25-mL Erlenmeyer flask.

4. Add 0.5 g of granular, anhydrous sodium sulfate to remove water. Allow to sit for 10 min., swirling occasionally. The solution should become clear. If it is still cloudy, add an additional 0.5 g of drying agent, swirl and allow to sit an additional 10 min.

Distillation

1. The methylcyclohexenes will be redistilled. Be certain that all the glassware you will use is washed thoroughly, is clean and is dry; dry all the glassware in an oven (110°C) for 10 min.

2. Transfer the dry methylcyclohexenes with a Pasteur pipet to a clean, dry 25-mL round bottom flask; add two (2) boiling stones.

3. Place the distilling flask into the heating mantle and attach to the apparatus. Begin heating by controlling the Variac. Collect product boiling in the range 102 - 111°C in a clean, dry preweighed 25-mL round bottom receiving flask (5). Reweigh the flask and product (6) and by subtraction, determine the weight of methylcyclohexenes (7); record these weights to the nearest 0.01 g on the Report Sheet. Calculate the percentage yield (9).

4. Place the sample in a vial of appropriate size. Make a neat label which contains the name of the product, the boiling range, the yield in grams, the percentage yield and your name. Turn in the sample to your instructor.

Analysis (Optional)

1. Your instructor will indicate which of the following sections to complete.

2. Obtain an infrared spectrum of the product. Use sodium chloride plates as described in Appendix I. Compare your spectrum with the spectrum of starting material (Fig. 8.1) and that of the product (Fig. 8.2) (10). Submit your spectrum with your Report Sheet.

3. Obtain a gas chromatogram of your product (see Expt. 3 for discussion). [A Hewlett-Packard Model 5890 Chromatograph using a HP-1 crosslinked methyl silicone gum column (30 m × 0.53 mm × 2.65 μm film thickness) at a column temperature of 90°C showed that methylcyclohexenes elute in the following order: 3-methylcyclohexene (retention time 1.08 min., b. p. 104°C); 4-methylcyclohexene (retention time 1.15 min., b. p. 103°C); 1-methylcyclohexene (retention time 1.23 min., b. p. 111°C). 2-Methylcyclohexanol does not interfere (retention time 2.76 min., b. p. 165.5°C).] Record the results on the Report Sheet (11).

Tests for Unsaturation (Optional)

1. Place 5 drops of methylcyclohexenes in one test tube (100 x 13 mm) and 5 drops of cyclo-hexane in a second test tube. To each test tube add bromine solution dropwise, until the red color persists. Count the number of drops required in each case; record on the Report Sheet (12).

CAUTION!   *Bromine solutions are toxic and corrosive. Avoid breathing vapors; avoid any contact with skin. Wear gloves and dispense in the hood. Dispose of all solutions in a waste container for halogenated hydrocarbons.*

2. Place 5 drops of methylcyclohexenes in one test tube (100 × 13 mm) and 5 drops of cyclo-hexane in a second test tube. Add 6 drops of 1,2-dimethoxyethane to each test tube and mix by gently tapping each test tube with your finger. Add aqueous potassium permanganate

solution to each test tube, with shaking. Observe which solution forms a brown precipitate; record your results on the Report Sheet (12). Dispose of your solutions in a waste container provided by your instructor.

CHEMICALS AND EQUIPMENT

1. Cyclohexane
2. 1, 2-Dimethoxyethane
3. 2-Methylcyclohexanol
4. Bromine solution, Br_2 in cyclohexane
5. 85% phosphoric acid, H_3PO_4
6. Potassium permanganate, aqueous, $KMnO_4$
7. 10% sodium carbonate solution, Na_2CO_3
8. Sodium sulfate, anhydrous, granular, Na_2SO_4
9. Boiling stones
10. Heating mantle
11. Latex tubing
12. Pasteur pipets
13. Ring clamp
14. 125-mL separatory funnel
15. Variac
16. Vial
17. Distillation kit
18. Optional: infrared spectrophotometer; sodium chloride discs and holder; gas chromatograph; 10- μL syringe

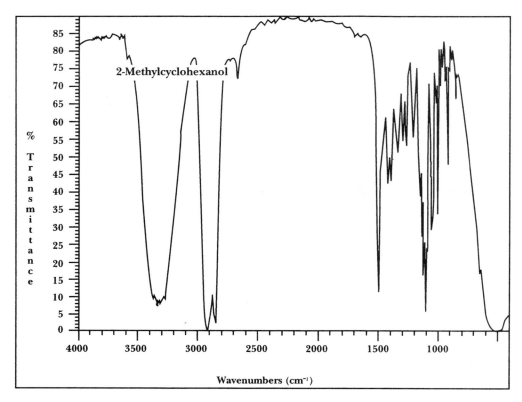

Figure 8.1 Infrared spectrum of 2-methylcyclohexanol, neat.

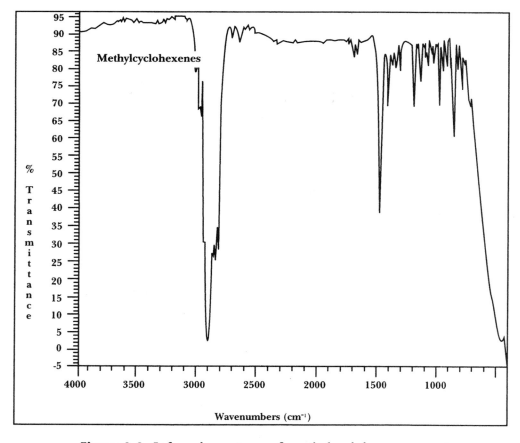

Figure 8.2 Infrared spectrum of methylcyclohexenes, neat.

Identification of Alcohols and Phenols

BACKGROUND

Specific groups of atoms in an organic molecule can determine its physical and chemical properties. These groups are referred to as *functional groups*. Organic compounds which contain the functional group −OH, the hydroxyl group, are called alcohols.

Alcohols are important commercially and include use as solvents, drugs and disinfectants. The most widely used alcohols are methanol or methyl alcohol, CH_3OH, ethanol or ethyl alcohol, CH_3CH_2OH, and 2-propanol or isopropyl alcohol, $(CH_3)_2CHOH$. Methyl alcohol is found in automotive products such as antifreeze and "dry gas." Ethyl alcohol is used as a solvent for drugs and chemicals, but is more popularly known for its effects as an alcoholic beverage. Isopropyl alcohol, also known as "rubbing alcohol," is an antiseptic.

Alcohols may be classified as either primary, secondary or tertiary:

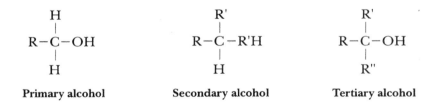

Primary alcohol   Secondary alcohol   Tertiary alcohol

Note that the classification depends on the number of carbon containing groups, R (alkyl or aromatic), attached to the carbon bearing the hydroxyl group. Examples of each type are as follows:

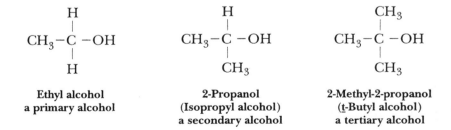

Ethyl alcohol
a primary alcohol

2-Propanol
(Isopropyl alcohol)
a secondary alcohol

2-Methyl-2-propanol
(t-Butyl alcohol)
a tertiary alcohol

Phenols bear a close resemblance to alcohols structurally since the hydroxyl group is present. However, since the −OH group is bonded directly to a carbon that is part of an aromatic ring, the chemistry is quite different from that of alcohols. Phenols are more acidic than alcohols; concentrated solutions of the compound phenol are quite toxic and can cause severe skin burns. Phenol derivatives are found in medicines; for example, thymol is used to kill fungi and hookworms. (Also see Table 9.1.)

TABLE 9.1 SELECTED ALCOHOLS AND PHENOLS

Compound	Name and Use
CH_3OH	Methanol: solvent for paints, shellacs and varnishes
CH_3CH_2OH	Ethanol: alcoholic beverages; solvent for medicines, perfumes and varnishes

$CH_3-CH-CH_3$
    |
    OH
	Isopropyl alcohol (2-propanol): rubbing alcohol; astringent; solvent for cosmetics, perfumes, and skin creams

CH_2-CH_2 | | OH OH	Ethylene glycol: antifreeze
$CH_2-CH-CH_2$ | | | OH OH OH	Glycerol (glycerin): sweetening agent; solvent for medicines; lubricant; moistening agent
	Phenol (carbolic acid): antiseptic
	Vanillin: flavoring agent (vanilla)
	Tetrahydrourushiol: irritant in poison ivy

Phenol

Thymol
(2-isopropyl-5-methylphenol)

In this laboratory, you will examine physical and chemical properties of representative alcohols and phenols. You will be able to compare the differences in chemical behavior between these compounds and use this information to identify an unknown.

Physical properties

Since the hydroxyl group is present in alcohols and phenols, these compounds are polar. The polarity of the hydroxyl group, coupled with its ability to form hydrogen bonds, enables alcohols and phenols to mix with water. Since these compounds also contain non-polar portions, they show additional solubility in many organic solvents, such as dichloromethane and diethyl ether.

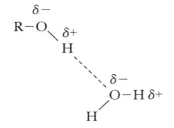

Hydrogen bonding of the hydroxyl group with water.

Chemical properties

The chemical behavior of the different classes of alcohols and of phenols can be used as a means of identification. Quick, simple tests that can be carried out in test tubes will be performed.

1. **Lucas test.** This test is used to distinguish between primary, secondary and tertiary alcohols. Lucas reagent is a mixture of zinc chloride, $ZnCl_2$, in concentrated HCl. Upon addition of this reagent a tertiary alcohol reacts rapidly and immediately gives an insoluble white layer. A secondary alcohol reacts slowly and after heating slightly gives the white layer within 10 min. A primary alcohol does not react. Any formation of a heterogeneous phase or appearance of an emulsion is a positive test.

$$CH_3CH_2-OH + HCl + ZnCl_2 \rightarrow \text{no reaction}$$
primary alcohol

$$(CH_3)_2CH-OH + HCl + ZnCl_2 \rightarrow (CH_3)_2CH-Cl \downarrow + H_2O \text{ (10 min., heat)}$$
secondary alcohol   insoluble

$$(CH_3)_3C-OH + HCl + ZnCl_2 \rightarrow (CH_3)_3C-Cl \downarrow + H_2O \text{ (immediate)}$$
tertiary alcohol   insoluble

2. **Chromic acid test.** This test is able to distinguish primary and secondary alcohols from tertiary alcohols. Using acidified dichromate solution primary alcohols are oxidized to aldehydes; secondary alcohols are oxidized to ketones; tertiary alcohols are not oxidized. (Note that in those alcohols which are oxidized, the carbon that has the hydroxyl group loses a hydrogen.) In the oxidation, the orange color of the chromic acid changes to a blue-green solution. Phenols are oxidized to nondescript brown tarry masses. (Aldehydes are also oxidized under these conditions, but ketones remain intact; see Experiment 14 for further discussion.)

$$3\ CH_3CH_2-OH + 4\ H_2CrO_4 + 6\ H_2SO_4 \rightarrow 3\ CH_3-\overset{\overset{O}{\|}}{C}-OH + 2\ Cr_2(SO_4)_3 + 13\ H_2O$$

   primary alcohol   **orange**   **carboxylic acid**   **blue-green**

$$3\ CH_3-\overset{\overset{OH}{|}}{C}H-CH_3 + 2\ H_2CrO_4 + 3\ H_2SO_4 \rightarrow 3\ CH_3-\overset{\overset{O}{\|}}{C}-CH_3 + 2\ Cr_2(SO_4)_3 + 8\ H_2O$$

  secondary alcohol   **orange**   **ketone**   **blue-green**

$$(CH_3)_3C-OH + H_2CrO_4 + H_2SO_4 \rightarrow \text{no reaction}$$

  tertiary alcohol

3. **Iodoform test.** This test is more specific than the previous two tests. Only ethyl alcohol and alcohols with the part structure $CH_3CH(OH)$ react. These alcohols react with iodine in aqueous sodium hydroxide to give the yellow precipitate iodoform.

$$\overset{\overset{OH}{|}}{R}CHCH_3 + 4\ I_2 + 6\ NaOH \rightarrow R\overset{\overset{O}{\|}}{C}O^-Na^+ + 5\ NaI + 5\ H_2O + HCl_{3(s)}$$

                                                                    iodoform
                                                                    yellow

Phenols also react under these conditions. With phenol, the yellow precipitate triiodophenol forms.

triiodophenol yellow

4. **Acidity of phenol.** Phenol is also called carbolic acid. Phenol is an acid and will react with base; thus, phenols readily dissolve in base solutions. In contrast, alcohols are not acidic.

5. **Ferric chloride test.** Addition of aqueous ferric chloride to a phenol gives a colored solution. Depending on the structure of the phenol, the color can vary from green to purple.

                             light yellow   **violet color**

OBJECTIVES

1. To learn characteristic chemical reactions of alcohols and phenols.
2. To use these chemical characteristics for identification of an organic compound.

PROCEDURE

CAUTION!   *Chromic acid is very corrosive. Any spill should be immediately flushed with water. Phenol is toxic and will cause burns to skin. Any contact should be thoroughly washed with large quantities of water. If phenol is dispensed as a solid, handle the solid only with a spatula or forceps. Use gloves with these reagents. Dispose of reaction mixtures and excess reagents in proper containers as directed by your instructor.*

Physical properties of alcohols and phenols

1. You will test the alcohols 1-butanol (a primary alcohol), 2-butanol (a secondary alcohol), 2-methyl-2-propanol (a tertiary alcohol), and phenol; you will also have as an unknown one of these compounds (labeled A, B, C or D). As you run a test on a known, test the unknown at the same time for comparison. Note that the phenol will be provided as an aqueous solution.

2. Into separate test tubes (100 x 13 mm) labeled 1-butanol, 2-butanol, 2-methyl-2-propanol and unknown place 10 drops of each sample; dilute by mixing with 3 mL of distilled water. Into a separate test tube place 2 mL of a prepared water solution of phenol. Are all the solutions homogeneous?

3. Test the pH of each of the aqueous solutions. Do the test by first dipping a clean glass rod into the solutions and then transferring a drop of liquid to pH paper. Use a broad indicator paper (e.g. pH range 1 – 12) and read the value of the pH by comparing the color to the chart on the dispenser. Record the results.

Chemical properties of alcohols and phenols

1. **Lucas test.** Place into separate clean, dry test tubes (100 × 13 mm) labeled 1-butanol, 2-butanol, 2-methyl-2-propanol, phenol and unknown 5 drops of each sample. Add 1 mL of Lucas reagent; mix well by stoppering each test tube with a cork and tap the test tube sharply with your finger for a few seconds to mix; remove the cork after mixing and allow each test tube to stand for 5 min. Look carefully for any cloudiness which may develop during this time period. If there is no cloudiness after 10 min., warm the test tubes that are clear for 15 min. in a 60°C water bath. Record your observations on the Report Sheet.

2. **Chromic acid test.** Place into separate clean, dry test tubes (100 × 13 mm), labeled as before, 5 drops of sample to be tested. To each test tube add 10 drops of reagent grade acetone and 2 drops chromic acid. Place the test tubes in a 60°C water bath for 5 min. Note the color of each solution. (Remember, the loss of the orange color and the formation of a blue-green color is a positive test.) Record your observations on the Report Sheet.

3. **Iodoform test.** Place into separate clean, dry test tubes (150 × 18 mm), labeled as before, 5 drops of sample to be tested. Add to each test tube, dropwise, 15 drops of 6 M NaOH; tap the test tube with your finger to mix. The mixture is warmed in a 60°C water bath, and the prepared solution of I₂–KI test reagent is added dropwise, with shaking, until the solution becomes brown (approx. 25 drops). Add 6 M NaOH, dropwise, until the solution becomes colorless. Keep the test tubes in the warm water bath for 5 min. Remove the test tubes from the water, let cool and look for a light yellow precipitate. Record your observations on the Report Sheet.

4. **Ferric chloride test.** Place into separate clean, dry test tubes (100 × 13 mm), labeled as before, 5 drops of sample to be tested. Add 2 drops of ferric chloride solution to each. Note any color changes in each solution. (Remember, a purple color indicates the presence of a phenol.) Record your observations on the Report Sheet.

CHEMICALS AND EQUIPMENT

1. Aqueous phenol
2. Acetone (reagent grade)
3. 1-Butanol
4. 2-Butanol
5. 2-Methyl-2-propanol (t-butyl alcohol)
6. Chromic acid solution
7. Ferric chloride solution
8. I₂-KI solution
9. Lucas reagent
10. Corks
11. Hot plate
12. pH paper
13. Unknown

Stereochemistry: Use of Molecular Models. II

EXPERIMENT 10

BACKGROUND

In Experiment 1 we looked at some molecular variations that acyclic organic molecules can take:

1. **Constitutional isomerism:** molecules can have the same molecular formula but different arrangements of atoms;

   a. **skeletal isomerism:** structural isomers where differences are in the order in which atoms that make up the skeleton are connected; e. g. C_4H_{10}

   $$CH_3CH_2CH_2CH_3$$
   Butane

   $$CH_3-\overset{\overset{\displaystyle CH_3}{|}}{C}H-CH_3$$
   2-Methylpropane

   b. **positional isomerism:** structural isomers where differences are in the location of a functional group; e. g. C_3H_7Cl

   $$CH_3CH_2CH_2-Cl$$
   1-Chloropropane

   $$CH_3-\overset{\overset{\displaystyle Cl}{|}}{C}H-CH_3$$
   2-Chloropropane

2. **Stereoisomerism:** molecules which have the same order of attachment of atoms, but differ in the arrangement of the atoms in 3-dimensional space.

   a. **cis/trans isomerism:** molecules that differ due to the geometry of substitution around a double bond; e. g. C_4H_8

   cis-2-Butene

   trans-2-Butene

   b. **conformational isomerism:** variation in acyclic molecules as a result of a rotation about a single bond; e. g. ethane, CH_3CH_3

Staggered   **Eclipsed**

In this experiment we will further investigate stereoisomerism by examining a cyclic system, cyclohexane, and several acyclic tetrahedral carbon systems. The latter possess more subtle characteristics as a result of the spatial arrangement of the component atoms. We will do this by building models of representative organic molecules and then studying their properties.

OBJECTIVES

1. To use models to study the conformations of cyclohexane.
2. To use models to distinguish between chiral and achiral systems.
3. To define and illustrate enantiomers, diastereomers and meso forms.
4. To learn how to represent these systems in 2-dimensional space.

PROCEDURE

You will build models and then you will be asked questions about the models. You will provide answers to these questions in the appropriate places on the Report Sheet. In doing this laboratory it will be convenient if you tear out the Report Sheet and keep it by the Procedure as you work through the exercises. In this way you can answer the questions without unnecessary turning back and forth.

Cyclohexane

Obtain a model set of "atoms" that contain the following:
    a. 8 carbon components – model atoms with 4-holes at the tetrahedral angle (e. g. black);
    b. 2 substituent components – model atoms with 1-hole (e. g. red);
    c. 18 hydrogen components – model atoms with 1-hole (optional) (e. g. white);
    d. 24 connecting links – bonds.

1. Construct a model of cyclohexane by connecting 6 carbon atoms in a ring; then into each remaining hole, insert a connecting link (bond), and if available, add a hydrogen to each.
    a. Is the ring rigid or flexible, that is, can the ring of atoms move and take various arrangements in space, or is the ring of atoms locked into only one configuration (1a)?
    b. Of the many configurations, which appears best for the ring: a planar or a puckered arrangement (1b)?
    c. Arrange the ring atoms into a chair conformation (Fig. 10.1a) and compare it to the picture of a lounge chair (Fig. 10.1b). (Does the term fit the picture?)

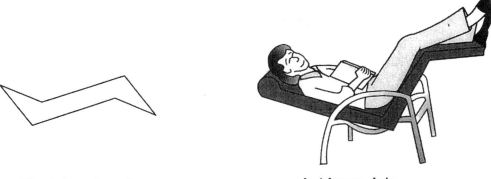

a. The chair conformation   b. A lounge chair

Figure 10.1 The chair conformation for a 6-carbon ring.

2. With the model in the chair conformation, rest it on the table top.
    a. How many hydrogens are in contact with the tabletop (2a)?
    b. How many hydrogens point in a direction 180° opposite to these (2b)?
    c. Take your pencil and place it into the center of the ring perpendicular to the table. Now rotate the ring around the pencil; we'll call this an axis of rotation. How many hydrogens are on bonds parallel to this axis (2c)? These hydrogens are called the axial hydrogens and the bonds are called the axial bonds.
    d. If you look at the perimeter of the cyclohexane system, the remaining hydrogens lie roughly in a ring perpendicular to the axis through the center of the molecule. How many hydrogens are on bonds lying in this ring (2d)? These hydrogens are called equatorial hydrogens and the bonds are called the equatorial bonds.
    e. Compare your model to the diagrams in Fig. 10.2 and be sure you are able to recognize and distinguish between axial and equatorial positions. In the space provided on the Report Sheet (2e), draw the structure of cyclohexane in the chair conformation with all 12 hydrogens attached. Label all the axial hydrogens H_a and all the equatorial hydrogens H_e. How many hydrogens are labeled H_a (2f)? How many hydrogens are labeled H_e (2g)?

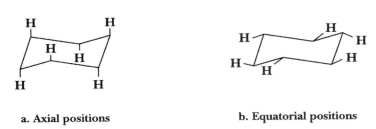

a. Axial positions   b. Equatorial positions

Figure 10.2 Axial and equatorial hydrogens in the chair conformation.

3. Look along any bond connecting any two carbon atoms in the ring. (Rotate the ring and look along a new pair of carbon atoms.) How are the bonds connected to these two carbons arranged? Are they staggered or are they eclipsed (3a)? In the space provided on the Report Sheet (3b), draw the Newman projection for the view (see Expt. 1 for an explanation of this projection); for the bond connecting a ring carbon, label that group "ring".

4. Pick up the cyclohexane model and view it from the side of the chair. Visualize the "ring" around the perimeter of the system perpendicular to the axis through the center. Of the 12 hydrogens, how many are pointed "up" relative to the plane (4a)? How many are pointed "down" (4b)?

5. Orient your model so that you look at an edge of the ring and it conforms to Fig. 10.3. Are the two axial positions labeled *A* cis

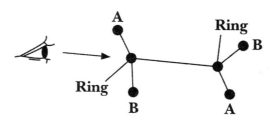

Figure 10.3 Cyclohexane ring viewed on edge.

or trans to each other (5a)? Are the two equatorial positions labeled *B* cis or trans to each other (5b)? Are the axial and equatorial positions *A* and *B* cis or trans to each other (5c)? Rotate the ring and view new pairs of carbons in the same way. See whether the relationship of positions vary from the above. Position your eye as in Fig. 10.3 and view along the carbon - carbon bond. In the space provided on the Report Sheet (5d), draw the Newman projection. Using this projection review your answers to 5a, 5b, and 5c.

6. Replace one of the axial hydrogens with a colored component atom. Do a "ring flip" by moving one of the carbons *up* and moving the carbon furthest away from it *down* (Fig. 10.4). In what position is the colored component after the ring flip (6a): axial or equatorial? Do an other ring flip. In what position is the colored component now (6b)? Observe all the axial positions and follow them through a ring flip.

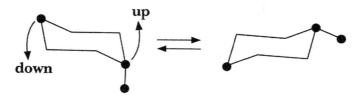

Figure 10.4 A "ring flip".

7. Refer to Fig. 10.3 and replace both positions labeled A by colored component atoms. Are they cis or trans (7a)? Do a ring flip. Are the two colored components cis or trans (7b)? Does the geometry change for the two components as the ring undergoes a ring flip (7c)? Repeat the exercise replacing atoms in positions labeled *A* and *B* and answer the same three questions for this model.

8. Replace one of the colored components with a methyl, $-CH_3$, group. Manipulate the model so that the $-CH_3$ group is in an axial position; examine the model. Do a ring flip placing the $-CH_3$ in an equatorial position; examine the model. Which of the chair conformations, $-CH_3$ axial or $-CH_3$ equatorial, is more crowded (8a)? What would account for one of the conformations being more crowded than the other (8b)? Which would be of higher energy, and thus, less stable (8c)? In the space provided on the Report Sheet (8d), draw the two conformations and connect with equilibrium arrows. Given your answers to 8a, 8b, and 8c, towards which conformation will the equilibrium lie (indicate by drawing one arrow bigger and thicker than the other)?

9. *A substituent group in the equatorial position of a chair conformation is more stable than the same substituent group in the axial position.* Do you agree or disagree? Explain your answer (9).

For the exercises in 10 - 15, although we will not be asking you to draw each and every conformation, we encourage you to practice drawing them in order to gain experience and facility in creating drawings on paper. Your instructor may make these exercises optional.

10. Construct *trans*-1,2-dimethylcyclohexane. By means of ring flips examine the model with the two $-CH_3$ groups axial and the two $-CH_3$ groups equatorial. Which is the more stable conformation? Explain your answer (10)?

11. Construct *cis*-1,2-dimethylcyclohexane by placing one $-CH_3$ group axial and the other equatorial. Do ring flips and examine the two chair conformations. Which is the more stable conformation? Explain your answer (11a). Given the two isomers, *trans*-1,2-dimethylcyclohexane and *cis*-1,2-dimethylcyclohexane, which is the more stable isomer? Explain your answer (11b).

12. Construct *cis*-1,3-dimethylcyclohexane by placing both $-CH_3$ groups in the axial positions. Do ring flips and examine the two chair conformations. Which is the more stable conformation? Explain your answer (12).

13. Construct *trans*-1,3-dimethylcyclohexane by placing one $-CH_3$ group axial and the other equatorial. Do ring flips and examine the two chair conformations. Which is the more stable conformation? Explain your answer (13a). Given the two isomers, *trans*-1,3-dimethylcyclohexane and *cis*-1,3-dimethylcyclohexane, which is the more stable isomer? Explain your answer (13b).

14. Construct *trans*-1,4-dimethylcyclohexane by placing both $-CH_3$ groups axial. Do ring flips and examine the two chair conformations. Which is the more stable conformation? Explain your answer (14).

15. Construct *cis*-1,4-dimethylcyclohexane by placing one $-CH_3$ group axial and the other equatorial. Do ring flips and examine the two chair conformations. Which is the more stable conformation? Explain your answer (15a). Given the two isomers, trans-1,4-dimethylcyclohexane and *cis*-1,4-dimethylcyclohexane, which is the more stable isomer? Explain your answer (15b).

16. Before we leave the cyclohexane ring system there are some additional ring conformations we can examine. As we move from one cyclohexane chair conformation to another, the *boat* is a transitional conformation between them (Fig. 10.5). Examine a model of the boat conformation by viewing along a carbon - carbon bond, as shown by Fig. 10.5. In the space provided on the Report Sheet (16a), draw the Newman projection for this view and compare with the Newman projection of 5d. By examining the models and comparing the Newman projections, explain which conformation, the chair or the boat, is more stable (16b). Replace the "flagpole" hydrogens by $-CH_3$ groups. What happens when this is done (16c)? The steric strain can be relieved by twisting the ring and separating the two bulky groups. What results is a *twist boat*.

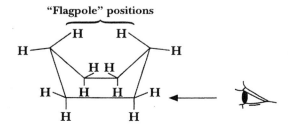

Figure 10.5 The boat conformation.

17. Review the conformations the cyclohexane ring can assume as it moves from one chair conformation to another:

chair ⇌ twist boat ⇌ boat ⇌ twist boat ⇌ chair

Chiral molecules

For this exercise obtain a model set of "atoms" which contain the following:

a. 8 carbon components - model atoms with 4-holes at the tetrahedral angle (e. g. black);
b. 32 substituent components - model atoms with 1-hole in four colors (e. g. 8 red; 8 white; 8 blue; 8 green or any other colors which your set may have);
c. 28 connecting links - bonds.

Also obtain a small hand mirror.

Enantiomers

1. Construct a model consisting of a tetrahedral carbon center with 4 different component atoms attached: red, white, blue, green; each color represents a *different* group or atom attached to carbon. Does this model have a *plane of symmetry* (1a)? A plane of symmetry can be described as a cutting plane — a plane that when passed through a model or object *divides it into two equivalent halves*; the elements on one side of the plane is the exact reflection of the elements on the other side. If you are using a pencil to answer these questions, examine the pencil. Does it have a plane of symmetry (1b)?

2. Molecules without a plane of symmetry are chiral. In the model you constructed in 1, the tetrahedral carbon is the chiral center; the molecule is chiral. A simple test for a chiral center in a molecule is to look for a carbon center with four different atoms or groups attached to it; this molecule will have no plane of symmetry. On the Report Sheet (2) are three structures; label with an asterisk (*) the chiral center in each structure.

3. Now take the model you constructed in 1 and set it in front of a mirror. Construct the model of the image projected in the mirror. You now have two models. If one is the object, what is the other (3a)? Do either have a plane of symmetry (3b)? Are both chiral (3c)? Now try to superimpose one model onto the other, that is, to place one model on top of the other in such a way that all five elements (i. e. the colored atoms) fall exactly one-on-top-of-the-other. Can you superimpose one model onto the other (3d)? *Enantiomers* are two molecules that are related to each other such that they are *nonsuperimposable mirror images of each other*. Are the two models you have a pair of enantiomers (3e)?

4. Molecules with a plane of symmetry are *achiral*. Replace the blue substituent with a second green one. The model should now have three different substituents attached to the carbon. Does the model now have a plane of symmetry (4a)? Passing the cutting plane through the model, what colored elements does it cut in half (4b)? What is on the left and right half of the cutting plane (4c)? Set this model in front of the mirror. Construct the model of the image projected in the mirror. You now have two models - an object and its mirror image. Are these two models superimposable on each other (4d)? Are the two models representative of different molecules or identical molecules (4e)?

Each stereoisomer in a pair of enantiomers has the property of being able to rotate mono-chromatic plane-polarized light. The instrument chemists use to demonstrate this property is called a polarimeter (see your text for a further description of the instrument). A pure solution of a single one of the enantiomers (referred to as an *optical isomer*) can rotate the light in either a clockwise (dextrorotatory, +) or a counterclockwise (levorotatory, −) direction. Thus, those mol-ecules that are optically active possess a "handedness" or chirality. Achiral molecules are optically inactive and do not rotate the light.

Meso forms and diastereomers

5. With your models, construct a pair of enantiomers. From each of the models remove the same common element (e. g. the white component) and the connecting links (bonds). Reconnect the two central carbons by a bond. What you have constructed is the *meso form* of a molecule, such as meso-tartaric acid. How many chiral carbons are there in this compound (5a)?

$$\text{HOOC}-\underset{\underset{\text{OH}}{|}}{\text{C}_1\text{H}}-\underset{\underset{\text{OH}}{|}}{\text{C}_2\text{H}}-\text{COOH}$$

Tartaric acid

Is there a plane of symmetry (5b)? Is the molecule chiral or achiral (5c)?

6. In the space provided on the Report Sheet (6) use circles to indicate the four different groups for carbon C_1 and squares to indicate the four different groups for carbon C_2.

7. Project the model into a mirror and construct a model of the mirror image. Are these two models superimposable or nonsuperimposable (7a)? Are the models identical or different (7b)?

8. Now take one of the models you constructed in 7, and on one of the carbon centers, exchange any two colored component groups. Does the new model have a plane of symmetry (8a)? Is it chiral or achiral (8b)? How many chiral centers are present (8c)? Take this model and one of the models you constructed in 7 and see whether they are superimposable. Are the two models superimposable (8d)? Are the two models identical or different (8e)? Are the two models mirror images of each other (8f)?

Here we have a pair of molecular models, each with two chiral centers, that are not mirror images of each other. These two examples represent *diastereomers*, stereoisomers that are not relat-ed as mirror images.

9. Take the new model you constructed in 8 and project it into a mirror. Construct a model of the image in the mirror. Are the two models superimposable (9a)? What term describes the relationship of the two models (9b)?

Thus, if we let these three models represent different isomers of tartaric acid, then we find that there are three stereoisomers for tartaric acid: a meso form and a pair of enantiomers. A meso form with any one of the enantiomers of tartaric acid represents a pair of diastereomers. Although it may not be true for this compound because of the meso form, in general, if you have **n** chiral centers, there are 2^n stereoisomers possible (see POST-LAB question no. 3).

Drawing stereoisomers

This section will deal with conventions for representing these 3-dimensional systems in 2-dimensional space.

10. Construct models of a pair of enantiomers; use tetrahedral carbon and four different colored components for the four different groups: red, green, blue, white. Hold one of the models in the following way:
    a. Grasp the blue group with your fingers and rotate the model until the green and red groups are pointing towards you (Fig. 10.6a). (Use the model which has the green group on the left and the red group on the right.)
    b. Holding the model in this way has the blue and white groups pointing away from you.
    c. If we use a drawing that describes a bond pointing towards you as a wedge and a bond pointing away from you as a dashed-line, the model can be drawn as shown in Fig. 10.6b.

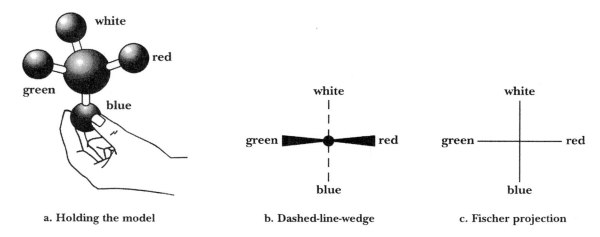

a. Holding the model   b. Dashed-line-wedge   c. Fischer projection

Figure 10.6 Projections in 2-dimensional space.

If this model were compressed into 2-dimensional space, we would get the projection shown in Fig. 10.6c. This is termed a Fischer projection and is named after a pioneer in stereochemistry, Emile Fischer. The Fischer projection has the following requirements:
1) the center of the cross represents the chiral carbon and is in the plane of the paper;
2) the horizontal line of the cross represents those bonds projecting out toward the the viewer from the plane of the paper;
3) the vertical line of the cross represents bonds projecting behind the plane of the paper.
d. In the space provided on the Report Sheet (10) use the enantiomer of the model in Fig. 10.6a and draw both the dashed-line-wedge and Fischer projection.

11. Take the model shown in Fig. 10.6a and rotate by 180°. Draw the Fischer projection (11a). Does this keep the requirements of the Fischer projection (11b)? Is the projection representative of the same system or of a different system (i. e. the enantiomer) (11c)?

In general, if you have a Fischer projection and rotate in the plane by 180°, the resulting projection is of the same system. Test this assumption by taking the Fischer projection in 10.6c, rotating 180° and comparing it to the drawing you did for 11a.

12. Again take the model shown in Fig. 10.6a. Exchange the red and the green components. Does this exchange give you the enantiomer (12a)? Now exchange the blue and the white components. Does this exchange return you to the original model (12b)?

In general, for a given chiral center, whether we use the dashed-line-wedge or the Fischer projection, an odd numbered exchange of groups leads to the mirror image of that center; an even numbered exchange of groups leads back to the original system.

13. Test the above by starting with the Fischer projection given below and carrying out the operations directed in a, b, and c; use the space provided on the Report Sheet (13) for the answers.

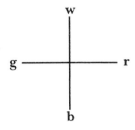

   a. Exchange *r* and *g*; draw the Fischer projection you obtain; label this new projection as either the same as the starting model or the enantiomer.

   b. Using the new Fischer projection from a., exchange *b* and *w*; draw the Fischer projection you now have.

   c. Now rotate the last Fischer projection you obtained by 180°; draw the Fischer projection you now have; label this as either the same as the starting model or the enantiomer.

14. Let us examine models with two chiral centers by using tartaric acid as the example, HOOC−CH(OH)−CH(OH)−COOH; use your colored components to represent the various groups. Hold your models so that each stereoisomer is oriented as in Fig. 10.7; in the space provided on the Report Sheet (14), draw each of the corresponding Fischer projections.

COOH	COOH	COOH
H►C◄OH	H►C◄OH HO►C◄H	
H►C◄OH	HO►C◄H H►C◄OH	
COOH	COOH	COOH
a. Meso	**b. Enantiomers**	

Figure 10.7 The stereoisomers of tartaric acid.

Circle the Fischer projection that shows a plane of symmetry. Underline all the Fischer projections that would be optically active.

15. Use the Fischer projection of meso-tartaric acid and on one chiral center carry out even and odd exchanges of the groups; follow these exchanges with a model. Does an odd exchange lead to an enantiomer, a diastereomer or to a system identical to the meso form (15a)? Does an even exchange lead to an enantiomer, a diastereomer or to a system identical to the meso form (15b)?

R/S convention for chiral centers

A system of assigning configuration to a chiral carbon was devised by the chemists R. S. Cahn, C. K. Ingold and V. Prelog. The system uses the letters *R* and *S* to designate the configuration at the chiral carbon: *R*, from the Latin *rectus*, or right; *S*, from the Latin *sinister*, or left. The designation arises from the priority order assigned to the four groups attached to the chiral carbon. Priority order is based on the atomic number of the atoms directly attached to the chiral carbon. The higher the atomic number, the higher the priority. In the case where two atoms have the same atomic number, you must move along the bonds to the next atoms out from the chiral center until an atom of different atomic number is reached. In order to apply the Cahn-Ingold-Prelog system, hold a model of the molecule so that the atom with *lowest* priority is pointed directly away from you. Then examine the remaining three attachments in terms of the order of their priorities (Fig. 10.8):

    a. If movement of your eye is from the highest to the lowest priority in a clockwise direction, the configuration of the chiral carbon is *R*.

    b. If movement of your eye is from the highest to the lowest priority in a counter-clockwise direction, the configuration of the chiral carbon is S.

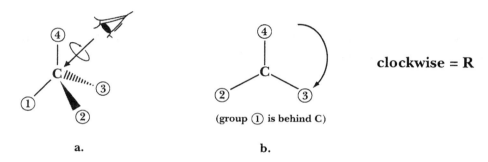

      a.   b.

Figure 10.8 An example of assignment using R/S priority rules: priority order 4 > 3 > 2 > 1.

Now let us work with a model to see how the rules apply.

16. Construct a model of 2-bromobutane, $CH_3CHBrCH_2CH_3$. (You should have enough carbon components for the carbon skeleton and colored components for H and Br.) Answer the following questions in the appropriate spaces on the Report Sheet.

    a. What are the four groups attached to the chiral carbon (16a)? Assign priorities to these groups: 4 = highest; 1 = lowest.

    b. View the model as shown in Fig. 10.8a and draw the arrangement you see. Is it *R* or *S* (16b)?

    c. Exchange the H and the Br. Again view the model and draw the arrangement. Is it *R* or *S* (16c)?

    d. Are the configurations opposite one another? Are they enantiomers (16d)?

CHEMICALS AND EQUIPMENT

Model kits vary in size and color of components. Use what is available; other colors may be substituted.

1. Cyclohexane model kit: 8 carbons (black, 4-holes); 18 hydrogens (white, 1-hole); 2 substituents (red, 1-hole); 24 bonds.

2. Chiral model kit: 8 carbons (black, 4-holes); 32 substituents (8 red, 1-hole; 8 white, 1-hole; 8 blue, 1-hole; 8 green, 1-hole); 28 bonds.

3. Hand mirror

EXPERIMENT 10

NAME _____ SECTION _____ DATE _____

PARTNER _____ GRADE _____

PRE-LAB QUESTIONS

1. Draw a chair conformation of cyclohexane which contains only axial hydrogens.

2. Look at your hands. Which term best explains the relationship of the two hands: identical, constitutional, conformational, enantiomers?

3. Define stereoisomers.

4. Label the chiral carbons in the molecules below with an asterisk (*).

$$CH_3-CH-CH-CH_3$$
with Cl and Cl attached to the two central CH carbons

5. Define optical isomer.

EXPERIMENT 10

NAME _____ SECTION _____ DATE _____

PARTNER _____ GRADE _____

<u>REPORT SHEET</u>

Cyclohexane

1. a.

    b.

2. a.

    b.

    c.

    d.

    e.

    f.

    g.

3. a.

    b.

4. a.

    b.

5. a.

    b.

    c.

    d.

6. a.

   b.

7.   <u>**Trial 1**</u>   <u>**Trial 2**</u>

   a.

   b.

   c.

8. a.

   b.

   c.

   d.

9.

10. e,e or a,a

11. a. a,e or e,a

    b.

12. a,a or e,e

13. a. a,e or e,a

    b.

14. a,a or e,e

15. a. a,e or e,a

   b.

16. a.

   b.

   c.

Enantiomers

1. a.
    b.

2.

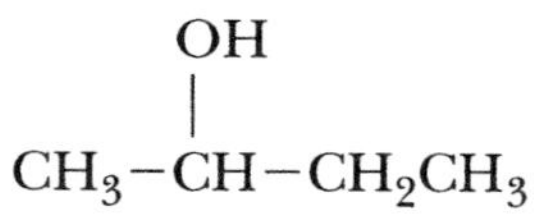

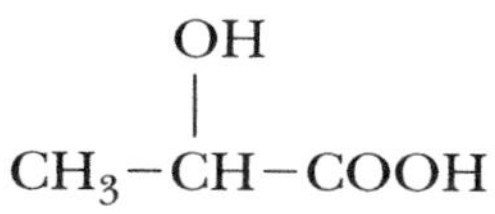

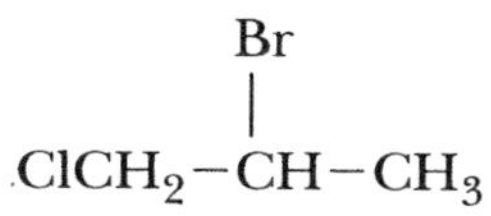

$$CH_3-\overset{\overset{\displaystyle OH}{|}}{CH}-CH_2CH_3 \qquad CH_3-\overset{\overset{\displaystyle OH}{|}}{CH}-COOH \qquad ClCH_2-\overset{\overset{\displaystyle Br}{|}}{CH}-CH_3$$

3. a.
    b.
    c.
    d.
    e.

4. a.
    b.
    c.
    d.
    e.

Meso forms and diastereomers

5. a.

   b.

   c.

6.

$$HOOC-\overset{\overset{\displaystyle H}{|}}{\underset{\underset{\displaystyle HO}{|}}{C_1}}-\overset{\overset{\displaystyle H}{|}}{\underset{\underset{\displaystyle OH}{|}}{C_2}}-COOH \qquad\qquad HOOC-\overset{\overset{\displaystyle H}{|}}{\underset{\underset{\displaystyle HO}{|}}{C_1}}-\overset{\overset{\displaystyle H}{|}}{\underset{\underset{\displaystyle OH}{|}}{C_2}}-COOH$$

7. a.

   b.

8. a.

   b.

   c.

   d.

   e.

   f.

9. a.

   b.

Drawing stereoisomers

10.

11. a.

   b.

   c.

12. a.

    b.

13. a.

    b.

    c.

14.

15. a.

    b.

R/S convention for chiral centers

16. a.

    b.

    c.

    d.

POST-LAB QUESTIONS

1. Draw the most stable conformation for *trans*-1-chloro-4-methylcyclohexane. Explain why you drew the molecule the way you did.

2. Draw the Fischer projections for the pair of enantiomers of lactic acid, $CH_3-CH(OH)-COOH$. Determine the configuration of each chiral carbon: R; S.

3. For 2,3-dibromopentane:
   a. How many stereoisomers are possible for this compound?

$$\begin{array}{cc} Br & Br \\ | & | \\ \end{array}$$
$$CH_3-CH-CH-CH_2CH_3$$

   b. Draw Fischer projections for each stereoisomer; label enantiomers. Label any meso isomer (if there are any).

   c. Assign R or S configurations for each chiral carbon in the stereoisomers.

4. Determine the relationship between the following pairs of structures: identical, enantiomers, diastereomers.

a.

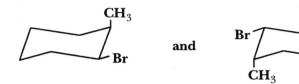

and

b.

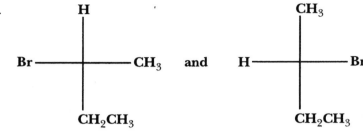

c.

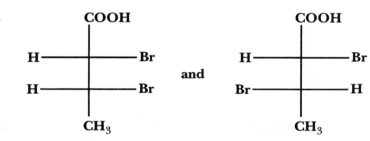

d.

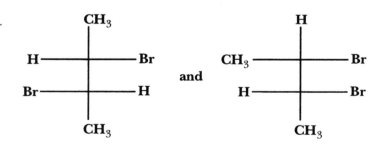

Reduction of α–Diketones by Yeast

BACKGROUND

It is often required that in an organic synthesis only one of the enantiomers of a chiral compound should be present as the product. The reduction of an α-diketone, such as 1-phenyl-1,2-propanedione is such an example. Reduction of the carbonyl groups on C1 and C2 can yield four stereoisomers:

(1)   (2)   (3)   (4)

In the determination of the R and S configuration of tetrahedral stereocenters, first the four different atoms attached to the chiral carbon (first shell atoms) are considered. Let's take 1-chloro-1-ethanol. The priorities among the first shell atoms follow their atomic numbers. Thus, $Cl>O>C>H$. Let's now orient the 1-chloro-1-ethanol molecule with its lowest priority branch ($-H$) away from the viewer.

$$H_3C \blacktriangleright \overset{\overset{\textstyle Cl}{|}}{\underset{\underset{\textstyle H}{\vdots}}{C}} \blacktriangleleft OH$$

(5)

If the direction going from the highest to the second highest to the third priority branch is a clockwise rotation the stereocenter is R configuration. If the route from highest to second highest to third priority branch follows a counter clockwise rotation the configuration is S. The example given above is, therefore, (R)-1-chloro-1-ethanol. In some cases the assignement of priority branch must extend beyond the first shell. Let us consider 2-butanol:

$$H_3C \underset{①}{\blacktriangleright} \overset{\overset{\textstyle OH}{|}}{\underset{\underset{\textstyle H}{\vdots}}{\underset{②}{C}}} \underset{③}{\blacktriangleleft CH_2}$$

(6)

Looking only at the first shell atoms, the priority would be $O > C_1 = C_3 > H$. In order to decide the priority between the C_1 and C_3 one must look at the atomic numbers in the second shell. C_1 has only H atoms, $(-CH_3)$, and C_3 has C and H atoms $(-CH_2-CH_3)$. Thus C_3 has a higher priority than C_1 and the total priority would be $O > C_3 > C_1 > H$. The molecule presented above (6) is called (R)-2-butanol. If needed, similar consideration can be extended to the third, etc., shells.

Using these rules in determination of the R and S configurations, we can assign the following names to the stereoisomers of 1-phenyl-1,2-propanediols represented by formulas (1) to (4):

(1) is (1R,2S); (2) is (1S,2S); (3) is (1R,2R) and (4) is (1S,2R).

When 1-phenyl-1,2-propanedione is reduced to the diol, depending on the experimental conditions, we obtain different mixtures of the stereoisomers (1) to (4). Stereospecificity can be achieved if we use enzymes which preferentially catalyze the production of mainly one stereoisomer. For example, yeast alcohol dehydrogenases preferentially produce (1R,2S) diols. In a yeast extract there are many enzymes, some of them also catalyzing the production of (1S,2S) configuration of 1-phenyl-1,2-propanediol. Fortunately, these enzymes are inhibited when the yeast extract is lyophilized (freeze dried). Therefore, in this experiment in which we would like to produce one stereoisomer only, we shall use freeze dried yeast as a catalyst in forming the desired product: (1R,2S)-1-phenyl-1,2-propanediol.

OBJECTIVES

1. Synthesis of a pure stereoisomer.
2. Characterization by gas and thin layer chromatography and infrared spectra.

PROCEDURE

This experiment is recommended to be performed in groups of two.

1. Add 0.5 L distilled water to a 1-L Erlenmeyer flask. Add a magnetic bar and place the flask on a magnetic stirrer. To the slowly swirling water add 40 g freeze dried baker's yeast, dispersing it evenly. Weigh approximately 500 mg 1-phenyl-1,2-propanedione. Record the weight to the nearest 0.001 g (1) and the number of moles (2) on your Report Sheet . Add the dione to the Erlenmeyer flask and continue to stir the mixture for 1 hr. at room temperature.

2. While waiting for the reaction to be completed, the characterization of the starting material, 1-phenyl-1,2-propanedione, will be accomplished by three different techniques. First, thin layer chromatography (TLC) will be performed. On a 10 × 4 cm TLC plate (silica gel on plastic sheet) mark a spot with pencil about 1 cm from the shorter edge (Fig.11.1). Use a plastic glove or hold the plate at the edges to avoid depositing your fingerprints on the TLC plate. Dip a capillary tube into the 0.1% 1-phenyl-1,2-propanedione solution (supplied by your stockroom) and apply it to the spot marked on the TLC plate. Do not let the spot spread beyond 1 mm in diameter. Let the spot dry. If a heat lamp is available, use it for drying. Repeat the spotting once more. Pour about 10 mL cyclohexane: diethyl ether mixture (50:50) into a 250-mL beaker. Place the spotted TLC plate into the beaker diagonally, making certain that the spot applied to the TLC plate is above the surface of the liquid. Cover the beaker with an aluminum foil to avoid the evaporation of the solvent mixture. Place the beaker in a hood. **CAUTION: Diethyl ether is highly flammable. No open flame, not even a hot-plate should be in the vicinity of its vapor.** Allow the solvent front to rise until it is 1-2 cm from the top

edge of the TLC plate (about 35 min.). *You must not allow the solvent front to advance up to or beyond the edge of the plate.* Remove the TLC plate from the beaker. Immediately mark with a pencil the position of the solvent front (Fig.11.1). Cover the beaker containing the eluents with an aluminum foil to use later (step No.8 of the Procedure). Allow the plate to dry in the hood. (You may facilitate the drying process using a heat lamp). Spray the TLC plate with aniline solution, dry it in an oven at 120°C for 10 min. Examine the TLC plate for a blue-green spot (a UV lamp, 254 nm, may help to locate the spot) and mark the spot with a pencil. Record on your Report Sheet the distance (in mm) the solvent front advanced (3). Record on your Report Sheet the distance your sample spot advanced (4). Calculate the R_f value of the dione and record it (5). (For reference see the Background section of Experiment 5).

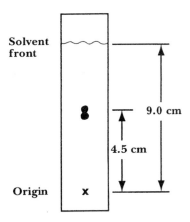

Figure 11.1 TLC chromatogram.

3. While the ascending TLC chromatography is in progress, establish the retention time of the starting compound, 1-phenyl-1,2-propanedione, by gas-liquid partition chromatography (GLPC). Many different GLPC equipments are available and, therefore, no specific instructions are given here for the use of the particular apparatus in your lab. (See Experiment 3)

    Introduce 1 µL of the 1% 1-phenyl-1,2-propanedione at the injection port. The temperature should be set for 190°C. Scanning time should be 15 min. Within that period you should notice on the chart printed out in your GLPC apparatus one main peak. Note its retention time, in minutes on your Report Sheet (6) and the area %, i.e. the relative concentration (7).

4. While still waiting for the completion of the TLC chromatogram and the GLPC you may do the characterization of your starting material by infrared spectroscopy. Alternatively, you may do this experiment at the end of the lab period when the spectrum of reactant, as well as that of the product, may be taken. The infrared spectrum characterizes the motion of certain atomic groups. For example, the stretching motion of the carbonyl groups, $-C=O$, have absorption bands at 1680 and 1720 cm^{-1}. In contrast the stretching motion of the alcohol, $-OH$, yields a broad absorption band from 3700-3100 cm^{-1}. (See Appendix 1 for further discussion).

    A 1% dispersion of 1-phenyl-1,2-propanedione in Nujol (mineral oil) is provided by your stockroom. Place a drop of this dispersion on a NaCl plate (window) and place a second NaCl plate to cover. Insert the two windows into a demountable sample holder consisting of two neoprene gaskets and two metal plates (Fig.11.2). Tighten the screws and insert the assembled IR cell into the IR spectrophotometer. (See also Appendix 1). Scan the spectrum from 4000 to 1000 cm^{-1}. Report the number of major absorption bands in the 1650-1750 cm^{-1} range and their actual wavenumbers (in cm^{-1}) on your Report Sheet (8,9).

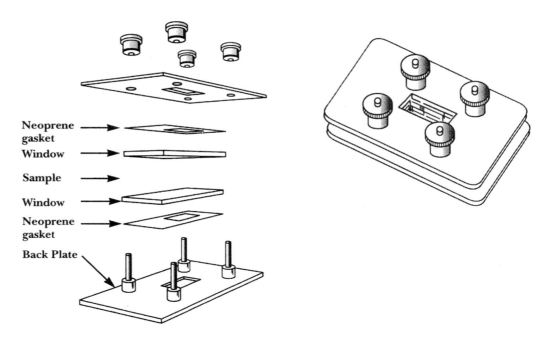

Figure 11.2 The assembly of an IR cell.

5. After 1 hr. stirring the reaction mixture the product is isolated and purified. Disperse about 5 g of Celite filter aid in 150 mL water. Insert a large Büchner funnel (85 mm OD) through a filtervac (or a neoprene adapter) into a 1-L filter flask and connect the side arm to a water aspirator with a heavy wall vacuum rubber tubing. A filter paper is placed on the Büchner filter and is wet with a few drops of distilled water. Make sure that the filter paper covers all the holes and lies flat. Turn on the water aspirator and pour the Celite slurry into the Büchner funnel. (Fig.11. 3).

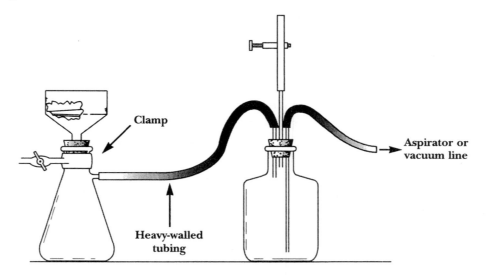

Figure 11.3 Vacuum filtration through Büchner funnel.

Allow the Celite to dry building up a filter cake on the filter paper. Remove the vacuum by disconnecting the rubber tubing from the side arm. Shut off the water aspirator. Remove the Büchner funnel. Discard the filtrate, wash the filter flask and rinse it with distilled water. Reconnect the side arm of the filter flask to the aspirator. Insert the Büchner funnel with filtervac into the filter flask and turn on the aspirator. Pour the reaction mixture into the

Büchner funnel. Wash the reaction flask first with 50 mL distilled water and then with 50 mL diethyl ether, and add both to the Büchner funnel. When the filtration is complete, remove the rubber tubing from the side arm and shut off the aspirator. The filtrate in the filter flask contains your product.

6. To isolate your product, add 150 g NaCl to the filtrate to saturate it. The product, 1-phenyl-1,2-propanediol, will be extracted from the filtrate using three portions of 60 mL diethyl ether. **CAUTION: Diethyl ether is highly flammable. The extraction should be carried out under the hood. No open flames, not even hot hot-plates should be in the vicinity.** (Alternatively, dichloromethane could be used instead of diethyl ether, but the three portions of extracting solvent must be increased to 150 mL. Under these conditions the yield will be smaller.) The successive extractions must be done with a separatory funnel, carefully venting the separatory funnel to avoid build up of vapor pressure (Fig.11.4). (See also Fig. 6.1, pg. 59, for technique). If during the extraction an emulsion is formed, transfer the emulsion to an Erlenmeyer flask and slowly with constant stirring add solid MgSO$_4$. Collect the organic layers in an Erlenmeyer flask; discard the aqueous layer. Dry the combined organic solvents (diethyl ether or dichloromethane) by adding anhydrous MgSO$_4$. Transfer the combined dehydrated extracts into a drying dish and remove the solvent by placing it on a steam bath under the hood.

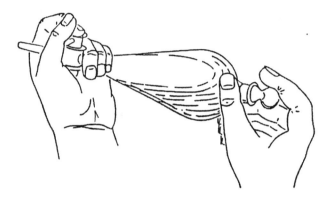

Figure 11.4 Using a separatory funnel.

7. The diol will be purified by gel chromatography. Place into a 25-mL buret a small piece of glass wool, deposited above the stopcock. Make a slurry of 3 g of silica gel and 10 mL dichloromethane. Add the slurry to the buret through a funnel and allow the silica gel to settle in the buret. Open the stopcock and drain the dichloromethane until it reaches the silica gel, but **do not let the column run dry**. Dissolve the dried diol (step No. 6) in a minimum amount of dichloromethane and deposit it on top of the column. Allow the diol to enter the column by opening the stopcock, but again make it certain that **the column does not run dry**. Fractionation is accomplished by adding solvents to the top of the column and draining it by opening the stopcock. The first fraction is obtained with 20 mL dichloromethane; the second with 3 mL diethyl ether and the third, which contains the pure diol, with 15 mL diethyl ether. Collect this last fraction in a preweighed evaporating dish, the weight of which you recorded on your Report Sheet to the nearest 0.001 g(10). Evaporate the solvent under vacuum in the hood and reweigh the evaporating dish. Report the weight on your Report Sheet (11). Calculate the yield of your reaction and report it on the Report Sheet (12, 13, 14).

8. Repeat the TLC procedures described under step 2 by using a drop (0.05 mL) of your purified diol dissolved in 0.5 mL of cyclohexane. Record the the distance of the solvent front (15) of the sample spot (16) and the calculated R$_f$ value (17) of your product, (1R,2S)-1-phenyl-1,2-propanediol on your Report Sheet.

9. To establish the purity of your product repeat the procedures described in step 3 using GLPC. Inject 1 μL of your product dissolved in cyclohexane into the porthole of your GLPC apparatus. Collect the data on the retention time of your product on the column at 190°C; run for 15 min. Report the retention times of your major peaks (18) and the % area (19) on your Report Sheet.

10. Finally to establish the identity of your product obtain an infrared spectrum as was described in step 4. Disperse a small amounts of your product in Nujol to give approximately 1% solution. Obtain the IR spectrum from 4000 to 1000 cm^{-1}. Report the number of major bands in the 3700-3100 cm^{-1} region, as well as in the 1650-1750 cm^{-1} range (20), and their peaks (in cm^{-1}) in the spectrum (20) on your Report Sheet.

CHEMICALS AND EQUIPMENT

1. 1-phenyl-1,2-propanedione
2. Freeze-dried baker's yeast
3. 0.1% 1-phenyl-1,2-propanedione in cyclohexane
4. Cyclohexane: diethyl ether, 50:50 mixture
5. 0.5% vanillin spray
6. 1% 1-phenyl-1,2-propanedione in cyclohexane
7. 1% 1-phenyl-1,2-propanedione dispersed in Nujol
8. Celite filter aid
9. NaCl, sodium chloride, granular
10. Diethyl ether
11. Magnesium sulfate, anhydrous
12. Dichloromethane
13. Silica gel
14. Cyclohexane
15. Nujol (mineral oil)
16. 10 × 4 cm silica gel TLC plates
17. Ruler
18. Polyethylene gloves
19. Capillary tubes open on both ends
20. Heat lamp or hair dryer
21. Drying oven, 120°C
22. UV lamp (254 nm)
23. Büchner funnel
24. 1-L filter flask
25. Filter paper
26. Filtervac or neoprene adaptor
27. 25-mL buret
28. Glass wool
29. Evaporating dish
30. Gas-liquid partition chromatography apparatus
31. Infrared spectrophotometer
32. Separatory funnel

EXPERIMENT 11

NAME _____ SECTION _____ DATE _____

PARTNER _____ GRADE _____

PRE-LAB QUESTIONS

1. Write the R and S configurations of bromochloroiodomethane.

2. Is 1-phenyl-1,2-propanedione a chiral compound? Explain.

3. What would be the reduction products of 2,3-pentanedione?

EXPERIMENT 11

NAME _____ SECTION _____ DATE _____

PARTNER _____ GRADE _____

REPORT SHEET

1. Weight of 1-phenyl-1,2-propanedione   _____ g

2. Moles of propanedione $[(1)/148.17]$ _____ moles

3. Distance the solvent front travelled   _____ mm

4. Distance the propanedione travelled   _____ mm

5. R_f value of propanedione $[(4)/(3)]$ _____

6. Retention time of propanedione on GLPC   _____ min

7. Area % of propanedione on GLPC   _____ %

8. Number of absorption bands of dione   _____

9. The peaks of the absorption bands in cm^{-1} _____

10. Weight of the evaporating dish (Tare)   _____ g

11. Weight of tare and product   _____ g

12. Weight of product $[(11) - (10)]$ _____ g

13. Moles of product $[(12)/150.19]$ _____ moles

14. Percent yield $[(13)/(2) \times 100]$ _____ %

15. Distance the solvent front travelled   _____ mm

16. Distance the propanediol spot travelled   _____ mm

17. R_f value of propanediol $[(16)/(15)]$ _____

18. Retention times of products on GLPC   _____ min.

                                                                    _____ min.

19. Area % of products on GLPC   _____ %

                                                                    _____ %

20. Number of major IR bands of products   _____

21. The peaks of major absorption bands in cm^{-1} _____

POST-LAB QUESTIONS

1. What was the role of anhydrous $MgSO_4$ in your isolation procedure? How would the percent yield be affected if you would omit this step?

2. Compare the results of the TLC and GLPC experiments of the reactant to that of the product.
    (a) Was your reactant a pure compound? Explain.

    (b) Were there more than one compound in the product?

    (c) Was there any evidence that your products still contained some 1-phenyl-1,2-propane-dione? Explain.

3. Does your IR spectrum support your answer given in 2(c)? Explain.

Preparation of 1-Bromobutane (n-Butyl bromide)

BACKGROUND

The functional group interconversion of an alcohol into an alkyl halide takes place by a nucleophilic substitution reaction, designated S_N. If the alkyl group is substituted on the tertiary carbon, the reaction proceeds by a two-step S_N1 mechanism: 1) the substrate dissociates to give a stable tertiary carbocation; 2) the nucleophile attacks the carbocation to give product. If the alkyl group is substituted on the primary carbon, the reaction proceeds by a concerted S_N2 mechanism: the nucleophile attacks the less crowded primary carbon center and gives product in one-step. (Refer to your text for a more comprehensive discussion.) The reaction in this experiment, the conversion of 1-butanol (n-butyl alcohol) into 1-bromobutane (n-butyl bromide), is an example of an S_N2 mechanism.

$$CH_3CH_2CH_2CH_2OH + NaBr + H_2SO_4 \longrightarrow CH_3CH_2CH_2CH_2Br + NaHSO_4 + H_2O$$
1-Butanol   **1-Bromobutane**

The reaction proceeds easily when the primary alcohol, 1-butanol, is reacted with a mixture of sodium bromide and concentrated sulfuric acid. The sulfuric acid protonates the hydroxyl group of the 1-butanol. The nucleophile, bromide, then attacks the primary carbon by an S_N2 mechanism. In the S_N2 process neutral water is a better leaving group than the hydroxyl anion.

$$CH_3CH_2CH_2CH_2-OH + H^+ \rightleftharpoons CH_3CH_2CH_2CH_2-\overset{\overset{\displaystyle H}{|}}{\underset{\underset{\displaystyle H}{|}}{O}}{}^+ \qquad \textbf{Fast first step}$$

$$CH_3CH_2CH_2CH_2-\overset{+}{O} \quad + Br^- \longrightarrow \left[\begin{array}{c} H \qquad H \\ \delta^- \diagdown \quad \diagup \delta^+ \diagup H \\ Br \text{---} C \text{---} O \\ | \qquad\quad H \\ CH_2CH_2CH_3 \end{array} \right] \longrightarrow CH_3CH_2CH_2CH_2-Br + H_2O$$

$$S_N2 \qquad\qquad\qquad\qquad\qquad\qquad \textbf{Slow second step}$$

The water that is produced in the reaction is then protonated (H_3O^+). This deactivates it and keeps it from acting as a nucleophile and reversing the reaction. Hydrobromic acid also is formed by the reaction of the sulfuric acid with the sodium bromide.

OBJECTIVES

1. To illustrate the preparation of 1-bromobutane by nucleophilic substitution.
2. To determine purity by gas chromatography.
3. To identify by infrared spectroscopy.

PROCEDURE

Reflux

1. Weigh a 50-mL round bottom flask to the nearest 0.01 g and record on the Report Sheet (1). Add 10.0 mL of 1-butanol (density = 0.810 g/mL; MW = 74.12 g/mol). Reweigh (2) and by subtraction determine the weight of the alcohol to 0.01 g (3); record these weights on the Report Sheet. Add to the flask 13.5 g of sodium bromide and 15 mL of water.

2. Cool the mixture in an ice-water bath and add 11.5 mL of concentrated sulfuric acid, by Pasteur pipet, slowly, with swirling and cooling. Add two (2) boiling chips to the flask.

CAUTION!   *Concentrated sulfuric acid will cause severe burns to the skin. Handle this chemical with care. If any acid should be spilled on your skin or your clothing, flush the area with plenty of water for 15 min. Dispense in the hood and wear gloves when using this reagent.*

3. The flask is then equipped with a vertically mounted water-cooled reflux condenser (Fig. 12.1). Heat with a heating mantle (controlled by a Variac) to a brisk reflux and continue for a period of 30 min. The heating mantle should be elevated so that it can be removed without disturbing the flask and the condenser. Note the reflux assembly has an inverted funnel above the reflux condenser connected by latex tubing to a water aspirator. (Any hydrogen bromide gas evolved during the reaction period will be drawn off and mixed with the water.)

Distillation

1. At the end of the reflux period, remove the heat source and let the mixture cool only until you can handle the round bottom flask comfortably; do not disconnect the reflux condenser while the system is cooling and do not allow the flask to cool to room temperature. (The 1-bromobutane is the upper layer since the aqueous solution of inorganic salts has a higher density.) Disconnect the flask from the reflux condenser and arrange for distillation (see Fig. 3.2, pg 24.). Heat with the heating mantle (controlled by a Variac).

2. Distill and collect the distillate in a 25-mL receiving flask until no more water-insoluble material comes over. Periodically you can determine this by carefully removing the receiving flask and collecting a few drops of distillate in a test tube; then add a few drops of water to the liquid collected in the test tube. If the liquid is water soluble, then all the 1-bromobutane has distilled and the distillation can be ended. (The 1-bromobutane distills along with water and appears cloudy.)

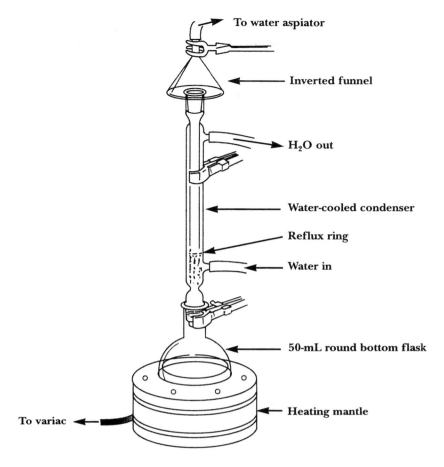

Figure 12.1 Reflux assembly.

Isolation

1. Pour the distillate into a 125-mL separatory funnel, add 10 mL of water and shake; vent after shaking by opening the stopcock to release gases (see Fig. 6.1, pg. 59, for technique). Set the separatory funnel on a ring clamp (see Fig. 6.2, pg. 59) and allow the layers to separate. Note that the 1-bromobutane now makes up the lower layer in accordance with a specific gravity of 1.275. (If there is a pink coloration due to a trace of bromine in the lower layer, it can be discharged by adding a small quantity of sodium bisulfite and shaking again.)

2. Drain the lower layer of 1-bromobutane into a clean Erlenmeyer flask. Then clean and dry the separatory funnel and return the 1-bromobutane to it.

3. Cool 10 mL of 9 M sulfuric acid in an ice bath and add it to the funnel. Shake well, vent by opening the stopcock, return to the ring clamp and allow 5 min. for separation of the layers.

4. Drain the lower aqueous layer. Allow 5 min. for further drainage and separate again.

5. Wash the 1-bromobutane with 10 mL of 10% aqueous sodium carbonate, to remove traces of acid, by shaking the separatory funnel and venting, as before. Allow the layers to settle. Separate and be careful to save the proper layer. (Save all layers until you are sure which layer is the 1-bromobutane. If you have a doubt, simply add a few drops of water to the separatory funnel and observe to which layer the water mixes. The water will be miscible with the aqueous layer. Now, which layer is the 1-bromobutane?)

6. Dry the cloudy 1-bromobutane by adding 1 to 2 g of anhydrous sodium sulfate. Add in 0.5 g increments, with swirling and by warming gently (*do not allow to boil*) on a steam bath or on a hot plate, until the liquid clears. (The object is to add the smallest amount of drying agent that will remove any water present.)

7. Decant the dried liquid through a funnel containing a small loose plug of cotton into a clean, dry 25-mL distilling flask. Add a boiling stone and distill. Be sure all the glassware for the distillation has been thoroughly cleaned and dried in an oven (110°C). Heat with a heating mantle (controlled by a Variac) and collect in separate receiving flasks the forerun and the product. (The distillate collected before the boiling range is the forerun.) The product is the material boiling in the range 99 - 103°C and should be collected in a preweighed 10-mL round bottom receiving flask (5). Measure with a graduated cylinder the volumes of forerun (9), product (10) and residue (11); record on the Report Sheet.

8. Reweigh the flask and product (6) and by subtraction determine the weight of 1-bromobutane (7); report the weights to the nearest 0.01 g on the Report Sheet. Calculate the percent yield (12).

9. Place the sample in a vial of appropriate size. Make a neat label which contains the name and formula of the product, the boiling range, the yield in grams, the percent yield and your name. Turn in the sample to your instructor.

10. *Waste Disposal.* All aqueous material may be flushed down the sink with plenty of water.

Analysis (Optional)

1. Your instructor will indicate which of the following parts to complete.

2. Determine the infrared spectrum of the product using sodium chloride discs. (Refer to Appendix I for a discussion of infrared spectroscopy.) Compare your spectrum to that of the 1-butanol (Fig. 12.2) and 1-bromobutane (Fig. 12.3) (13). Submit your spectrum with your Report Sheet.

3. Obtain a gas chromatogram of your product (see Expt. 3 for discussion). Report the percent composition of the product sample on the Report Sheet (14). [A Hewlett-Packard Model 5890 Chromatograph using a HP-1 crosslinked methyl silicone gum column (30 m × 0.53 mm × 2.65 μm film thickness) at a column temperature of 75°C showed retention times for 1-butanol of 1.3 min. and for 1-bromobutane of 1.8 min.]

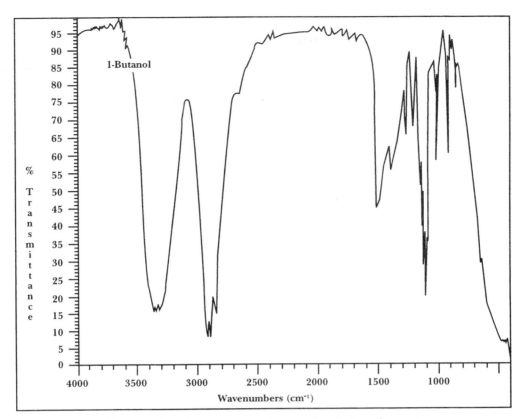

Figure 12.2 Infrared spectrum of 1-butanol, neat.

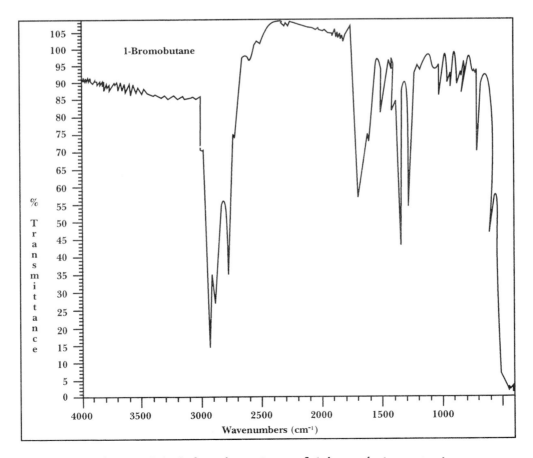

Figure 12.3 Infrared spectrum of 1-bromobutane, neat.

CHEMICALS AND EQUIPMENT

1. 1-Butanol (n-butyl alcohol), $CH_3CH_2CH_2CH_2OH$
2. Sodium bisulfite, $NaHSO_3$
3. Sodium bromide, NaBr
4. 10% sodium carbonate, aqueous, Na_2CO_3
5. Sodium sulfate, anhydrous, granular, Na_2SO_4
6. Concentrated sulfuric acid, H_2SO_4
7. 9 M sulfuric acid, H_2SO_4
8. Boiling chips
9. Funnel, glass
10. Heating mantle
11. Latex tubing
12. Pasteur pipets
13. Ring clamp
14. 125-mL separatory funnel
15. Variac
16. Vial
17. Reflux/distillation kit
18. Optional equipment: infrared spectrophotometer; sodium chloride discs and holder; gas chromatograph; 10-μmL syringe

EXPERIMENT 12

NAME _____ SECTION _____ DATE _____

PARTNER _____ GRADE _____

PRE-LAB QUESTIONS

1. Define an S_N2 mechanism.

2. Why should 1-butanol be expected to react by an S_N2 mechanism?

3. What is the purpose of the sulfuric acid in the reaction?

4. What is the nucleophile in the reaction for this experiment?

5. Complete the equations below:

   a. $CH_3CH_2CH_2CH_2CH_2OH + NaBr + H_2SO_4 \longrightarrow$

   b. $CH_3CH_2CH_2CH_2OH + NaI + H_2SO_4 \longrightarrow$

EXPERIMENT 12

NAME _____ SECTION _____ DATE _____

PARTNER _____ GRADE _____

REPORT SHEET

1. Weight of 50-mL Round bottom flask _____ g
2. Weight of 50-mL Round bottom flask plus alcohol _____ g
3. Weight of 1-butanol (2) − (1) _____ g
4. Moles of 1-butanol [(3)/74.12] _____ moles

5. Weight of 10-mL Round bottom flask _____ g
6. Weight of 10-mL Round bottom flask plus product _____ g
7. Weight of 1-bromobutane (6) - (5) _____ g
8. Moles of 1-bromobutane [(7)/137.02] _____ moles

9. Forerun _____ mL
10. Product _____ mL
11. Residue _____ mL

12. % Yield $= \dfrac{\text{moles of 1-bromobutane (8)}}{\text{moles of 1-butanol (4)}} \times 100$

13. Comparison of spectra

14. Composition by gas chromatography

RETENTION TIME, MIN.	IDENTITY	% COMPOSITION

POST-LAB QUESTIONS

1. In the isolation procedure, step no. 3, 9 M H_2SO_4 is added. Write an equation that shows how this is an effective reagent for removal of unreacted 1-butanol.

2. Possible by-products in this reaction are an ether and an alkene. Draw structures for these two products.

3. At the end of the reflux period, if the solution is allowed to cool to room temperature, salts precipitate. What is the likely composition of the salts?

4. When 1-bromobutane and 10% aqueous sodium carbonate are combined (isolation step no. 5), two layers form. Which is the aqueous layer? Explain your answer.

5. In determining percent yield, the number of moles of 1-butanol is used in the calculation. Why can you use this value instead of grams?

6. What is the advantage of carrying out a reflux procedure for 30 min. *versus* boiling in an Erlenmeyer flask for the same period of time?

A Friedel-Crafts Alkylation

EXPERIMENT 13

BACKGROUND

The Friedel-Crafts alkylation of aromatic rings (arenes) usually involves an alkyl halide with a Lewis acid catalyst.

$$\text{Arene} + \text{R–X} \xrightarrow[\text{cat.}]{\text{Lewis acid}} \text{Alkyl arene} + \text{HX}$$

Arene   Alkyl halide   Alkyl arene

In order for the reaction to work, the aromatic ring should be unsubstituted or should have an activating group attached (e. g. $-OH$, $-OCH_3$, $-CH_3$); deactivating groups (e. g. $-NO_2$) generally fail to give Friedel-Crafts products. In this reaction, p-xylene (1,4-dimethylbenzene) is used. However, since alkyl groups activate and the alkylated aromatic ring is more reactive than the starting material, multiple substitution can result. In order to minimize multiple alkylations, the starting arene usually is present in excess.

Another difficulty in carrying out an alkylation reaction is that alkyl halides, with the exception of methyl and ethyl, can lead to alkyl arenes different from those expected because of rearrangements. In particular, primary alkyl halides rearrange under the reaction conditions to secondary or tertiary carbocations. In this experiment, the primary alkyl halide, n-propyl chloride, is the reactant. As a result, the propyl p-xylene products which form contain either the unrearranged n-propyl group or the rearranged isopropyl group.

$$\text{1,4-Dimethylbenzene (p-Xylene)} + CH_3CH_2CH_2Cl + AlCl_3 \longrightarrow \text{1,4-Dimethyl-2-n-propylbenzene} + \text{1,4-Dimethyl-2-isopropylbenzene}$$

1,4-Dimethylbenzene (p-Xylene)   1-Chloropropane (n-Propyl chloride)   1,4-Dimethyl-2-n-propylbenzene   1,4-Dimethyl-2-isopropylbenzene

133

Here, the attacking species is either the initial Friedel-Crafts complex or the secondary, isopropyl cation formed by a hydride migration that accompanies ionization of the carbon-chlorine bond.

$$\underset{\textbf{Friedel-Crafts complex}}{CH_3-\overset{\overset{H}{|}}{CH}-CH_2-\overset{+}{Cl}-\overset{-}{AlCl_3}} \longrightarrow \underset{\textbf{Isopropyl cation}}{CH_3-\overset{+}{CH}-\overset{\overset{H}{|}}{CH_2}} + \overset{-}{AlCl_4}$$

Gas chromatography will be used to determine the product distribution.

The Lewis acid catalyst that will be used is aluminum chloride. This is a strong catalyst, and as a catalyst, is used in non-stoichiometric quantities. Other halides that can be used are those of boron (BF_3), tungsten ($SbCl_3$), iron ($FeCl_3$), tin ($SnCl_4$), and zinc ($ZnCl_2$).

OBJECTIVES

1. To carry out a Friedel-Crafts alkylation of an activated benzene ring.
2. To show that rearrangement of a primary alkyl group occurs.
3. To determine the percentage composition of unrearranged and rearranged alkylated product.

PROCEDURE

Reaction

1. Weigh to the nearest 0.01 g a clean, dry 25-mL round bottom flask and record on the Report Sheet (1). Add 7.4 mL (density = 0.8611 g/mL; MW = 106.16) of dry p-xylene; reweigh (2) and by subtraction, determine the weight of p-xylene (3); record to the nearest 0.01 g on the Report Sheet.

2. Add a small magnetic spin-bar to the flask. Attach a Claisen adapter. To one of the openings of the Claisen adapter attach a drying tube containing anhydrous calcium chloride; seal the other opening with a rubber septum cap (Fig. 13.1).

3. Weigh out 0.30 g of anhydrous aluminum chloride (MW=133.34). Record the weight of aluminum chloride to the nearest 0.01 g on the Report Sheet (5). (**CAUTION! Do this in the hood; aluminum chloride is toxic; the dust is hygroscopic and irritating. The aluminum chloride *must be weighed out quickly* in order to avoid reaction with atmospheric moisture. Clean up spilled material immediately.**) Add immediately to the reaction flask by separating the Claisen adapter from the flask and pouring the aluminum chloride through the neck opening. Quickly seal. Stir the contents with a magnetic stirrer at room temperature.

4. Weigh to the nearest 0.01 g a clean, dry 10-mL Erlenmeyer flask; record on the Report Sheet (7). Add 2.7 mL (density = 0.8923 g/mL; MW = 78.54) of n-propyl chloride (1-chloropropane); reweigh (8) and by subtraction, determine the weight of n-propyl chloride (9); record to the nearest 0.01 g on the Report Sheet. Transfer the n-propyl chloride to the reaction mixture with a syringe by inserting the syringe needle through the rubber septum and adding the n-propyl chloride dropwise.

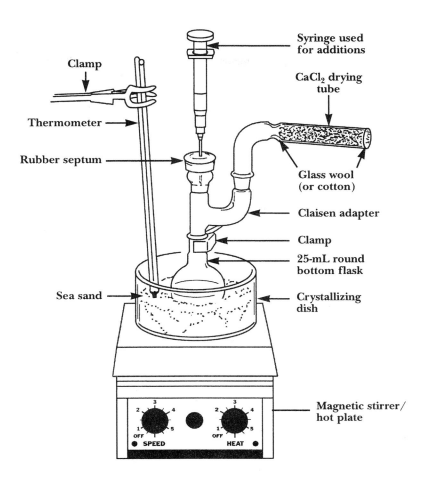

Figure 13.1 Friedel-Crafts apparatus.

5. The reaction mixture is placed in a sand bath (use a 125 × 65 mm Pyrex crystallizing dish and a layer of sand deep enough to heat the mixture). Bring to 50°C and heat and stir at that temperature for one hr.

6. After heating add 8.0 mL of water to the stirred solution; keep the solution stirred until the aluminum chloride is consumed. Transfer the solution to a 125-mL separatory funnel supported by a ring clamp (see Fig. 6.2, pg 59. for technique) and draw off the lower aqueous layer; this may be discarded.

7. Wash the organic solution again with 6.0 mL of 5% aqueous sodium bicarbonate solution; remove the aqueous layer and discard. Repeat with another 6.0 mL of water. After removal of the last aqueous wash, the organic solution is poured into a clean, dry 25-mL Erlenmeyer flask. Add 2.0 g of anhydrous sodium sulfate to remove any remaining water; allow to sit for 10 min., swirling occasionally.

8. The solution is decanted or pipetted with a Pasteur pipet from the drying agent into a clean, dry vial of appropriate size.

Analysis

1. Obtain a gas chromatogram of your product (see Expt. 3 for discussion). [A Hewlett-Packard Model 5890 Chromatograph using a HP-5 crosslinked phenyl methyl silicone gum column (30 m × 0.25 mm × 0.25 μm film thickness) at a column temperature of 120°C showed p-xylene eluting with a retention time of 1.7 min., 1,4-dimethyl-2-isopropylbenzene eluting with a retention time of 3.3 min., and 1,4-dimethyl-2-n-propylbenzene eluting with a retention time of 3.6 min.] Record your results on the Report Sheet (11).

CHEMICALS AND EQUIPMENT

1. Aluminum chloride, anhydrous, $AlCl_3$
2. 5% aqueous sodium bicarbonate, $NaHCO_3$
3. Calcium chloride, anhydrous, $CaCl_2$
4. n-Propyl chloride (1-chloropropane)
5. Sea Sand
6. Sodium sulfate, granular, anhydrous, Na_2SO_4
7. p-Xylene
8. Calcium chloride drying tube
9. Claisen adapter
10. Crystallizing dish, Pyrex
11. Magnetic spin-bar
12. Magnetic stirrer/hot plate
13. Pasteur pipets
14. Ring clamp
15. 25-mL round bottom flask
16. Rubber septum
17. 125-mL separatory funnel
18. Syringe, 1-mL
19. Syringe, 10-μL
20. Thermometer, 110°C
21. Vial

EXPERIMENT 13

NAME _____   SECTION _____   DATE _____

PARTNER _____   GRADE _____

PRE-LAB QUESTIONS

1. Is the preparation in this laboratory the best method for obtaining 1,4-dimethyl-2-<u>n</u>-propyl-benzene? Explain your answer.

2. Write an equation for a Friedel-Crafts reaction that would give only 1,4-dimethyl-2-isopropyl-benzene as the product.

3. Why is it necessary to be cautious in working with aluminum chloride?

4. What is the purpose of the aluminum chloride in the procedure?

EXPERIMENT 13

NAME _____ SECTION _____ DATE _____

PARTNER _____ GRADE _____

REPORT SHEET

1. Weight of 25-mL flask   _____ g
2. Weight of 25-mL flask and p-xylene   _____ g
3. Weight of p-xylene (2) − (1)   _____ g
4. Moles of p-xylene [(3)/106.16)]   _____ moles

5. Weight of aluminum chloride   _____ g
6. Moles of aluminum chloride [(5)/133.34)]   _____ moles

7. Weight of 10-mL flask   _____ g
8. Weight of 10-mL flask and n-propyl chloride   _____ g
9. Weight of n-propyl chloride (8) − (7)   _____ g
10. Moles of n-propyl chloride [(9)/78.54)]   _____ moles

11. Gas chromatography

RETENTION TIME, MIN.	IDENTITY	% COMPOSITION

POST-LAB QUESTIONS

1. Write an equation for the reaction of aluminum chloride with water.

2. In the procedure, step no. 6, you are asked to "...draw off the lower aqueous layer...."

    a. Why can you be sure that this is the aqueous layer?

    b. However, if you were not sure, devise a simple experiment that you could do that would identify the aqueous layer.

3. Why do you get only monoalkylation products in this Friedel-Crafts procedure?

4. What is the limiting reagent for the reaction in this procedure? Based on this determination, calculate the theoretical (or expected) yield of product in moles and in grams. (The MW of the product is 148.24.)

Properties of Amines and Amides

BACKGROUND

Two classes of organic compounds which contain nitrogen are amines and amides. Amines behave as organic bases and may be considered as derivatives of ammonia. Amides are compounds which have a carbonyl group connected to a nitrogen atom and are neutral. In this experiment you will learn about the physical and chemical properties of some members of the amine and amide families.

If the hydrogens of ammonia are replaced by alkyl or aryl groups, amines result. Depending on the number of organic groups attached to nitrogen, amines are classified as either primary (one group), secondary (two groups), or tertiary (three groups) (Table 14.1).

TABLE 14.1 TYPES OF AMINES

	Primary Amines	**Secondary Amines**	**Tertiary Amines**
NH_3 Ammonia	CH_3NH_2 Methylamine	$(CH_3)_2NH$ Dimethylamine	$(CH_3)_3N$ Trimethylamine
	Aniline	N-Methylaniline	N,N-Dimethylaniline

There are a number of similarities between ammonia and amines that carry beyond the structural. Consider odor. The smell of amines resembles that of ammonia, but are not as sharp. However, amines can be quite pungent. Anyone handling or working with raw fish knows how strong the amine odor can be since raw fish contains low-molecular weight amines such as dimethylamine and trimethylamine. Other amines associated with decaying flesh have names suggestive of their odors: putrescine and cadaverine.

$$NH_2CH_2CH_2CH_2CH_2NH_2 \qquad NH_2CH_2CH_2CH_2CH_2CH_2NH_2$$

Putrescine
(1,4-Diaminobutane)

Cadaverine
(1,5-Diaminopentane)

The solubility of low molecular weight amines in water is high. In general, if the total number of carbons attached to nitrogen is six or less, the amine is water soluble; amines with a carbon content greater than six are water insoluble. However, all amines are soluble in organic solvents such as diethyl ether or methylene chloride.

Since amines are organic bases, water solutions show weakly basic properties. If the basicity of aliphatic amines and aromatic amines are compared to ammonia, aliphatic amines are stronger than ammonia, while aromatic amines are weaker. Amines characteristically react with acids to form ammonium salts; the non-bonded electron pair on nitrogen bonds the hydrogen ion.

$$R\ddot{N}H_2 + HCl \longrightarrow RNH_3^+Cl^-$$

Amine **Ammonium Salt**

If an amine is insoluble, reaction with an acid produces a water-soluble salt. Since ammonium salts are water soluble, many drugs containing amines are prepared as ammonium salts. After working with fish in the kitchen, a convenient way to rid one's hands of fish odor is to rub a freshly cut lemon over the hands. The citric acid found in the lemon reacts with the amines found on the fish; a salt forms which can be easily rinsed away with water.

Amides are carboxylic acid derivatives. The amide group is recognized by the nitrogen connected to the carbonyl group. Amides are neutral compounds; the electrons are delocalized into the carbonyl (resonance).

Amide group   **Acetamide**   **Benzamide**

Under suitable conditions amide formation can take place between an amine and a carboxylic acid, an acyl halide or an acid anhydride. Along with ammonia, primary and secondary amines yield amides with carboxylic acids or derivatives. Table 14.2 relates the nitrogen base with the amide class (based on the number of alkyl or aryl groups on the nitrogen of the amide.)

$$CH_3NH_2 + CH_3COOH \longrightarrow CH_3COO^-(CH_3NH_3^+) \xrightarrow{\Delta} CH_3CONHCH_3 + H_2O$$

$$CH_3NH_2 + CH_3COCl \longrightarrow CH_3CONHCH_3 + HCl$$

TABLE 14.2 CLASSES OF AMIDES

Nitrogen Base	Amide (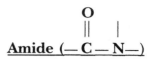)
Ammonia	Primary amide (no R groups)
Primary Amine	Secondary amide (one R group)
Secondary Amine	Tertiary amide (two R groups)

Hydrolysis of amides can take place in either acid or base. Primary amides hydrolyze in acid to ammonium salts and carboxylic acids. Neutralization of the acid and ammonium salts releases ammonia which can be detected by odor or by litmus.

$$\underset{\substack{\| \\ \text{R--C--NH}_2}}{\text{O}} + \text{HCl} + \text{H}_2\text{O} \longrightarrow \underset{\substack{\| \\ \text{R--C--OH}}}{\text{O}} + \text{NH}_4\text{Cl}$$

$$\text{NH}_4\text{Cl} + \text{NaOH} \longrightarrow \text{NH}_3 + \text{NaCl} + \text{H}_2\text{O}$$

Secondary and tertiary amides would release the corresponding alkyl ammonium salts which, when neutralized, would yield the amine.

In base primary amides hydrolyze to carboxylic acid salts and ammonia. The presence of ammonia (or amine from corresponding amides) can be detected similarly by odor or litmus. The carboxylic acid would be generated by neutralization with acid.

$$\underset{\substack{\| \\ \text{R--C--NH}_2}}{\text{O}} + \text{NaOH} \longrightarrow \underset{\substack{\| \\ \text{R--C--O}^-\text{Na}^+}}{\text{O}} + \text{NH}_3$$

$$\underset{\substack{\| \\ \text{R--C--O}^-\text{Na}^+}}{\text{O}} + \text{HCl} \longrightarrow \underset{\substack{\| \\ \text{R--C--OH}}}{\text{O}} + \text{NaCl}$$

OBJECTIVES

1. To show some physical and chemical properties of amines and amides.
2. To demonstrate the hydrolysis of amides.

PROCEDURE

CAUTION! *Amines are toxic chemicals. Avoid excessive inhaling of the vapors and direct skin contact. Anilines are more toxic and are readily absorbed through the skin. Wash any amine or aniline spill with large quantities of water. Use gloves with these chemicals.*

Properties of amines

1. Place 5 drops of liquid or 0.1 g of solid from the compounds listed in the table below into labeled clean, dry test tubes (100 × 13 mm).

Test Tube No.	Nitrogen Compound
1	6 M NH_3
2	Triethylamine
3	Aniline
4	N,N-Dimethylaniline
5	Acetamide

2. Carefully note the odors of each compound. Do not inhale deeply. Merely wave your hand across the mouth of the test tube toward your nose in order to note the odor. Record your observations on the Report Sheet.

3. Add 2 mL of distilled water to each of the labeled test tubes. Mix thoroughly by sharply tapping the test tube with your finger. Note on the Report Sheet whether the amines are soluble or insoluble.

4. Take a glass rod and test each solution for its pH. Carefully dip one end of the glass rod into a solution and touch a piece of pH paper. Between each test, be sure to clean and dry the glass rod. Record the pH by comparing the color of the paper with the chart on the dispenser.

5. Carefully add 2 mL of 6 M HCl to each test tube. Mix thoroughly by sharply tapping the test tube with your finger. Compare the odor and solubility of this solution to previous observations.

6. Place 5 drops of liquid or 0.1 g of solid from the compounds listed in the table into labeled clean, dry test tubes (100 × 13 mm). Add 2 mL of diethyl ether (ether) to each test tube. Stopper with a cork and mix thoroughly by shaking. Record the observed solubilities.

7. Carefully place on a watch glass, side-by-side, without touching, a drop of triethylamine and a drop of concentrated HCl. Record your observations.

Hydrolysis of acetamide

1. Dissolve 0.5 g of acetamide in 5 mL of 6 M H_2SO_4 in a large test tube (150 × 18 mm). Heat the solution in a boiling water bath for 5 min.

2. Hold a small strip of moist pH paper over the mouth of the test tube; note any changes in color; record the pH reading. Remove the test tube from the water bath holding it in a test tube holder. Carefully note any odor.

3. Place the test tube in an ice-water bath until cool to the touch. Now carefully add, dropwise, with shaking, 6 M NaOH to the cool solution until basic. (You will need more than 7 mL of base.) Hold a piece of moist pH paper over the mouth. Record the pH reading. Carefully note any odor.

CHEMICALS AND EQUIPMENT

1. Acetamide
2. 6 M NH_3, ammonia water
3. Aniline
4. N,N-Dimethylaniline
5. Triethylamine
6. Diethyl ether (ether)
7. 6 M NaOH
8. Concentrated HCl
9. 6 M HCl
10. 6 M H_2SO_4
11. pH papers
12. Hot plate

EXPERIMENT 14

NAME _____ SECTION _____ DATE _____

PARTNER _____ GRADE _____

PRE-LAB QUESTIONS

1. Draw the structure of the functional group that is found in an amine and an amide.

2. What is the general rule for the solubility of amines in water?

3. What happens when an amine is mixed with hydrochloric acid?

4. Compare the basicity of the following amines to ammonia: ethylamine (an aliphatic amine), aniline (an aromatic amine).

EXPERIMENT 14

NAME _____ SECTION _____ DATE _____

PARTNER _____ GRADE _____

REPORT SHEET

Properties of Amines

	Odor		Solubility			pH
	Original Sol.	with HCl	H_2O	Ether	HCl	H_2O
6 M NH_3						
Triethylamine						
Aniline						
N,N-Dimethylaniline						
Acetamide						

Triethylamine and concentrated hydrochloric acid observation:

Hydrolysis of acetamide

1. Acid solution

   a. pH reading:

   b. Odor noted:

2. Base solution

   a. pH reading:

   b. Odor noted:

POST-LAB QUESTIONS

1. The active ingredient in the commercial insect repellent "OFF" is N,N-diethyl-m-toluamide. Is this a primary, secondary or tertiary amide?

2. Of the amines tested which was the least soluble in water? Why?

3. Why does triethylamine lose its odor when mixed with hydrochloric acid?

4. Write the equations that account for what happens in the hydrolysis of the acetamide solution in a) acid and in b) base.

    a.

    b.

15

Identification of Aldehydes and Ketones

BACKGROUND

Aldehydes and ketones are representative of compounds which possess the carbonyl group:

$$\diagup C = O$$ **The carbonyl group.**

Aldehydes have at least one hydrogen attached to the carbonyl carbon; in ketones, no hydrogens are directly attached to the carbonyl carbon, only carbon containing R-groups:

R—C—H $\|$ O   R—C—R' $\|$ O   **(R and R' can be alkyl or aromatic)**

Aldehyde   **Ketone**

Aldehydes and ketones of low molecular weight have commercial importance. Many others occur naturally. Table 15.1 has some representative examples.

TABLE 15.1 REPRESENTATIVE ALDEHYDES AND KETONES

Compound		Source and Use
$\overset{O}{\overset{\|}{HCH}}$	Formaldehyde	Oxidation of methanol; plastics; preservative
$\overset{O}{\overset{\|}{CH_3CCH_3}}$	Acetone	Oxidation of isopropyl alcohol; solvent
	Citral	Lemon grass oil; fragrance
	Jasmone	Oil of jasmine; fragrance

In this experiment you will investigate the chemical properties of representative aldehydes and ketones.

Classification tests

1. **Chromic acid test.** Aldehydes are oxidized to carboxylic acids by chromic acid; ketones are not oxidized. A positive test is the formation of a blue-green solution from the orange color of chromic acid.

$$3R-\overset{\overset{\textstyle O}{\|}}{C}-H \; + \; 2H_2CrO_4 \; + \; 3H_2SO_4 \; \longrightarrow \; 3R-\overset{\overset{\textstyle O}{\|}}{C}-OH \; + \; Cr_4(SO_4)_3 + \; 5H_2O$$

aldehyde   orange   blue-green

$$R_2C = O \; \xrightarrow[\text{H}_2\text{SO}_4]{\text{H}_2\text{CrO}_4} \; \text{no reaction}$$

ketone

2. **Tollens' test.** Most aldehydes reduce Tollens' reagent (ammonia and silver nitrate) to give a precipitate of silver metal. The free silver forms a silver mirror on the sides of the test tube. (This test is sometimes referred to as the "silver mirror" test.) The aldehyde is oxidized to a carboxylic acid.

$$R\overset{\overset{\textstyle O}{\|}}{C}-H \; + \; 2\,Ag(NH_3)_2OH \; \longrightarrow \; 2\,Ag(s) \; + \; R\overset{\overset{\textstyle O}{\|}}{C}O^-NH_4^+ \; + \; H_2O \; + \; 3\,NH_3$$

aldehyde   silver mirror

3. **Iodoform test.** Methyl ketones give the yellow precipitate iodoform when reacted with iodine in aqueous sodium hydroxide.

$$R\overset{\overset{\textstyle O}{\|}}{C}-CH_3 \; + \; 3\,I_2 \; + \; 4\,NaOH \; \longrightarrow \; 3\,NaI \; + \; 3\,H_2O \; + \; R\overset{\overset{\textstyle O}{\|}}{C}O^-NH_4^+ \; + \; HCI_{3(s)}$$

methyl ketone   iodoform
yellow

4. **2,4-Dinitrophenylhydrazine test.** All aldehydes and ketones give an immediate precipitate with 2,4-dinitrophenylhydrazine reagent. This reaction is general for both these functional groups. The color of the precipitate varies from yellow to red. (Note that alcohols do not give this test.)

$$R-\overset{\overset{\textstyle H}{|}}{C}=O \; + \; H_2N-NH-\underset{}{\bigcirc}-NO_2 \; \to \; R-\overset{\overset{\textstyle H}{|}}{C}=N-NH-\underset{}{\bigcirc}-NO_2$$

aldehyde   yellow to red

$$R-\overset{\overset{\displaystyle R}{|}}{C}=O \ + \ H_2N-NH-\underset{}{\overset{\overset{\displaystyle NO_2}{|}}{\bigcirc}}-NO_2 \ \rightarrow \ R-\overset{\overset{\displaystyle R}{|}}{C}=N-NH-\underset{}{\overset{\overset{\displaystyle NO_2}{|}}{\bigcirc}}-NO_2$$

ketone   yellow to red

$$ROH \ + \ H_2N-NH-\underset{}{\overset{\overset{\displaystyle NO_2}{|}}{\bigcirc}}-NO_2 \ \rightarrow \ \text{no reaction}$$

TABLE 15.2 SUMMARY OF CLASSIFICATION TESTS

Compound	Reagent for Positive Test
Aldehydes and Ketones	2,4-Dinitrophenylhydrazine
Aldehydes	Chromic acid
	Tollens' reagent
Methyl Ketones	Iodoform

Identification by forming a derivative

The classification tests, when properly done, can distinguish between various types of aldehydes and ketones. However, these tests alone may not allow for the identification of a specific unknown aldehyde or ketone. A way to correctly identify an unknown compound is by using a known chemical reaction to convert it into another compound that is known. The new compound is referred to as a *derivative*. Then by comparing the physical properties of the unknown and the derivative to the physical properties of known compounds listed in a table, an identification can be made.

The ideal derivative is a solid. A solid can be easily purified by crystallization and easily characterized by its melting point. Thus, two similar aldehydes or two similar ketones usually have derivatives that have *different melting points*. The most frequently formed derivatives for aldehydes and ketones are the 2,4-dinitrophenylhydrazone (2,4-DNP), oxime and semicarbazone. Table 15.3 lists some aldehydes and ketones along with melting points of their derivatives. If, for example, we look at the properties of valeraldehyde and crotonaldehyde, though the boiling points are virtually the same, the melting points of the 2,4-DNP, oxime and semicarbazone are different and provide a basis for identification.

1. **2,4-Dinitrophenylhydrazone**. 2,4-Dinitrophenylhydrazine reacts with aldehydes and ketones to form 2,4-dinitrophenylhydrazones (2,4-DNP).

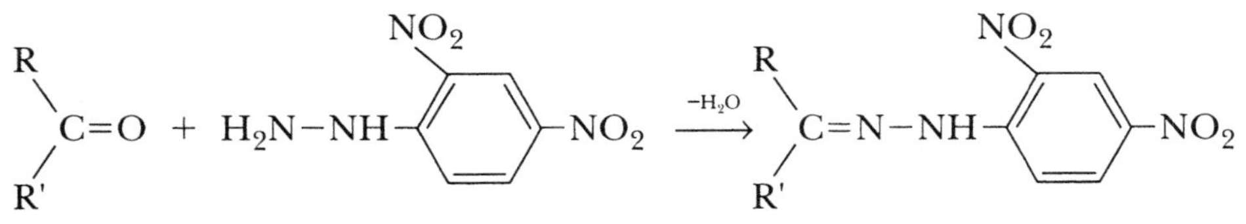

$$\underset{R'}{\overset{R}{\diagdown}}C=O \ + \ H_2N-NH-\underset{\underset{}{}}{\bigcirc}\overset{NO_2}{}-NO_2 \ \xrightarrow{-H_2O} \ \underset{R'}{\overset{R}{\diagdown}}C=N-NH-\underset{}{\bigcirc}\overset{NO_2}{}-NO_2$$

         2,4-dinitrophenylhydrazine   **2,4-dinitrophenylhydrazone (2,4-DNP)**

(R' = H or alkyl or aromatic)

2. The 2,4-DNP product is usually a colored solid (yellow to red) and is easily purified by recrystallization.

Oxime. Hydroxylamine reacts with aldehydes and ketones to form oximes.

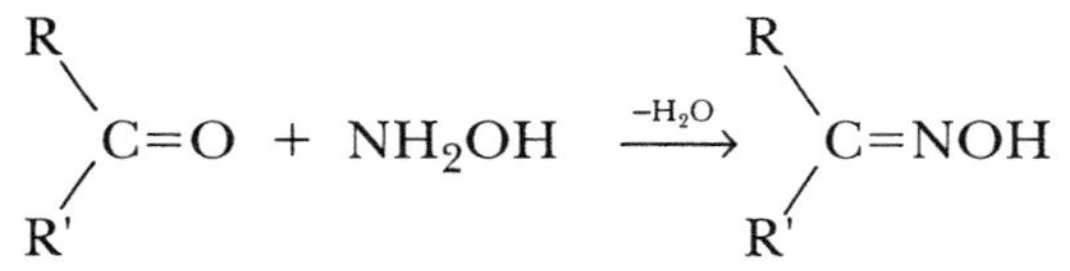

$$\underset{R'}{\overset{R}{\diagdown}}C=O \ + \ NH_2OH \ \xrightarrow{-H_2O} \ \underset{R'}{\overset{R}{\diagdown}}C=NOH$$

         hydroxylamine   **oxime**

(R' = H or alkyl or aromatic)

These are usually lower melting derivatives.

3. **Semicarbazone.** Semicarbazide, as its hydrochloride salt, reacts with aldehydes and ketones to form semicarbazones.

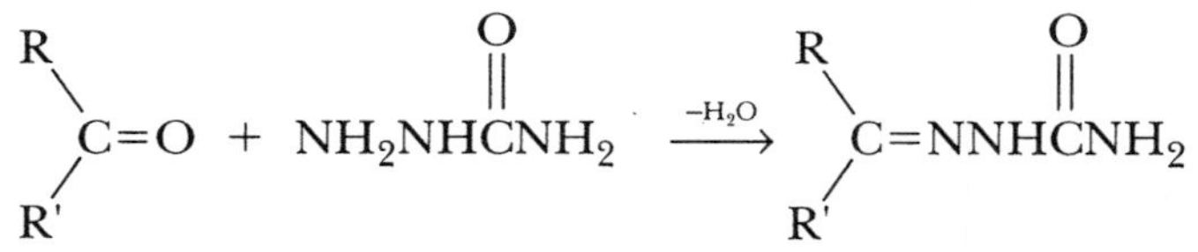

$$\underset{R'}{\overset{R}{\diagdown}}C=O \ + \ NH_2NH\overset{O}{\overset{\|}{C}}NH_2 \ \xrightarrow{-H_2O} \ \underset{R'}{\overset{R}{\diagdown}}C=NNH\overset{O}{\overset{\|}{C}}NH_2$$

         semicarbazide   **semicarbazone**

(R' = H or alkyl or aromatic)

A pyridine base is used to neutralize the hydrochloride in order to free the semicarbazide so it may react with the carbonyl substrate.

TABLE 15.3 SELECTION OF ALDEHYDES AND KETONES WITH DERIVATIVES[1]

Compound	Formula	b.p. (°C)	2,4-DNP m.p. (°C)	Oxime m.p. (°C)	Semicarbazone m.p. (°C)
Aldehydes					
Isovaleraldehyde (2-methylbutanal)	$CH_3-CH-CH_2-C=O$ (with CH_3 and H)	93	123	49	107
Valeraldehyde (pentanal)	$CH_3CH_2CH_2CH_2-C=O$ (with H)	103	106	52	—
Crotonaldehyde (2-butanal)	$CH_3-CH=CH-C=O$ (with H)	104	190	119	199
Caprylaldehyde (octanal)	$CH_3CH_2CH_2CH_2CH_2CH_2CH_2C=O$ (with H)	171	106	60	101
Benzaldehyde	$C_6H_5-C=O$ (with H)	178	237	35	222
Ketones					
Acetone (2-propanone)	$CH_3-C(=O)-CH_3$	56	126	59	187
2-Pentanone	$CH_3-C(=O)-CH_2CH_2CH_3$	102	144	58 (b.p. 167°C)	112
3-Pentanone	$CH_3CH_2-C(=O)-CH_2CH_3$	102	156	69 (b.p. 165°C)	139
Cyclopentanone	(cyclopentanone ring =O)	131	146	56	210
Cyclohexanone	(cyclohexanone ring =O)	156	162	90	166
Acetophenone	$C_6H_5-C(=O)-CH_3$	202	238	60	198

1.   CRC Handbook of Tables for Organic Compound Identification, 3 rd Ed., compiled by Zvi Rappoport, The Chemical Rubber Co., Cleveland (1967).

OBJECTIVES

1. To learn the chemical characteristics of aldehydes and ketones.
2. To use these chemical characteristics in simple tests to distinguish between examples of aldehydes and ketones.
3. To identify aldehydes and ketones by formation of derivatives.

PROCEDURE

Classification tests

1. Classification tests are to be carried out on four known compounds and one unknown. Any one test should be carried out on all five samples at the same time for comparison. Label test tubes as shown in Table 15.4.

TABLE 15.4 LABELING TEST TUBES

Test Tube No.	Compound
1	Isovaleraldehyde (an aliphatic aldehyde)
2	Benzaldehyde (an aromatic aldehyde)
3	Cyclohexanone (a ketone)
4	Acetone (a methyl ketone)
5	Unknown

CAUTION!   *Chromic acid is toxic and corrosive. Handle with care and wash promptly any spill. Use gloves with this reagent.*

2. **Chromic acid test.** Place 5 drops of each substance into separate, labeled test tubes (100 × 13 mm). Dissolve each compound in 20 drops of reagent grade acetone (to serve as solvent); then add to each test tube 4 drops of chromic acid reagent, one drop at a time; after each drop mix by sharply tapping the test tube with your finger. Let stand for 10 min. Aliphatic aldehydes should show a change within a minute; aromatic aldehydes take longer. Note the approximate time for any change in color or formation of a precipitate on the Report Sheet.

CAUTION!   *The Tollens' reagent must be freshly prepared before it is to be used and any excess disposed of immediately after use. Organic residues should be discarded in appropriate waste containers; unused Tollens' reagent should be flushed away in the sink with large quantities of water.* **Do not store Tollens' reagent since it is explosive when dry.**

3. **Tollens' test.** Enough reagent for your use can be prepared in a 25-mL Erlenmeyer flask by mixing 5 mL of Tollens' solution A with 5 mL of Tollens' solution B. To the silver oxide precipitate which forms, add dropwise, with shaking, 10% ammonia solution until the brown precipitate just dissolves. *Avoid an excess of ammonia.*

   Place 5 drops of each substance into separately labeled clean, dry test tubes (100 × 13 mm). Dissolve the compound in bis(2-ethoxyethyl)ether by adding this solvent dropwise until a homogeneous solution is obtained. Then add 2 mL (40 drops) of the prepared Tollens' reagent and mix by sharply tapping the test tube with your finger. Place the test tube in a 60°C water bath for 5 min. Remove the test tubes from the water and look for a silver mirror. Record your results on the Report Sheet.

4. **Iodoform test.** Place 10 drops of each substance into separately labeled clean, dry test tubes (150 × 18 mm). Add to each test tube, dropwise, with shaking, 25 drops of 6 M NaOH. The mixture is warmed in a water bath (60°C), and the prepared solution of I_2-KI test reagent is added dropwise, with shaking, until the solution becomes brown (approx. 35 drops). Add 6 M NaOH until the solution becomes colorless. Keep the test tubes in the warm water bath for 5 min. Remove the test tubes from the water, let cool and look for a light yellow precipitate. Record your observations on the Report Sheet.

5. **2,4-Dinitrophenylhydrazine test.** Place 5 drops of each substance into separately labeled clean, dry test tubes (100 × 13 mm) and add 20 drops of the 2,4-dinitrophenylhydrazine reagent to each. If no precipitate forms immediately heat for 5 min. in a warm water bath (60°C); cool. Record your observations on the Report Sheet.

Formation of derivatives

CAUTION!   *The chemicals used to prepare derivatives and some of the derivatives are potential carcinogens. Exercise care in using the reagents and in handling the derivatives. Avoid skin contact by wearing gloves.*

1. **General procedure for recrystallization.** Heat a small volume (10 - 20 mL) of solvent to boiling on a steam bath (or carefully on a hot plate). Place crystals into a test tube (100 × 13 mm) and add the hot solvent, dropwise, until the crystals just dissolve (keep the solution hot, also). Allow the solution to cool to room temperature; then cool further in an ice bath. Collect the crystals on a Hirsch funnel by vacuum filtration (use a trap between the Hirsch funnel set-up and the aspirator; Fig. 15.1); wash the crystals with 10 drops of ice cold solvent. Allow crystals to dry by drawing air through the Hirsch funnel. Take a melting point (see Experiment 4 for technique).

2.   Your instructor will indicate how many derivatives and which derivatives you should make.

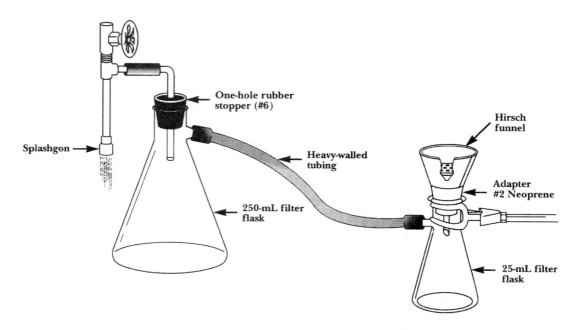

One-hole rubber
stopper (#6)

Hirsch
funnel

Splashgon →

Heavy-walled
tubing

Adapter
#2 Neoprene

250-mL filter
flask

25-mL filter
flask

Figure 15.1   Vacuum filtration with a Hirsch Funnel.

3.   **2,4-Dinitrophenylhydrazone.** Place 5 mL of the 2,4-dinitrophenylhydrazine reagent in a test tube (150 × 18 mm). Add 10 drops of the unknown compound; sharply tap the test tube with your finger to mix. If crystals do not form immediately, gently heat in a water bath (60°C) for 5 min. Then cool in an ice bath until crystals form. Collect the crystals by vacuum filtration using a Hirsch funnel (Fig. 15.1). Allow the crystals to dry on the Hirsch funnel by drawing air through the crystals. Take a melting point and record on the Report Sheet. (The crystals are usually pure enough to give a good melting point. However, if the melting point range is too large, recrystallize from a minimum volume of ethanol.)

4.   **Oxime.** Prepare fresh reagent by dissolving 1.0 g of hydroxylamine hydrochloride and 1.5 g of sodium acetate in 4 mL of distilled water in a test tube (150 × 18 mm). Add 20 drops of unknown and sharply tap the test tube with your finger to mix. Warm in a hot water bath (60°C) for 5 min. Cool in an ice bath until crystals form. (If no crystals form, scratch the inside of the test tube with a glass rod.) Collect the crystals on a Hirsch funnel by vacuum filtration (Fig. 15.1). Allow the crystals to air dry on the Hirsch funnel by drawing air through the crystals. Take a melting point and record on the Report Sheet. (Recrystallize, if necessary from a minimum volume of ethanol.)

5.   **Semicarbazone.** Place 2.0 mL of the semicarbazide reagent in a test tube (150 × 18 mm); add 10 drops of unknown. If the solution is not clear, add methanol dropwise until a clear solution results. Add 2.0 mL of pyridine and gently warm in a hot bath (60°C) for 5 min. Crystals should begin to form. (If there are no crystals, place the test tube in an ice bath and scratch the inside of the test tube with a glass rod.) Collect the crystals on a Hirsch funnel by vacuum filtration (Fig. 15.1). Allow the crystals to air dry on the Hirsch funnel by drawing air through the crystals. Take a melting point and record on the Report Sheet. (Recrystallize, if necessary, from a minimum volume of ethanol.)

6. *Waste.* Place all the waste solutions from these preparations in designated waste containers for disposal by your instructor.

7. Based on the observations you recorded on the Report Sheet, and by comparing the melting points of the derivatives for your unknown to the knowns listed in Table 15.3, identify your unknown.

CHEMICALS AND EQUIPMENT

1. Acetone (reagent grade)
2. 10% ammonia solution
3. Benzaldehyde
4. Bis(2-ethoxyethyl) ether
5. Chromic acid reagent
6. Cyclohexane
7. 2,4-dinitrophenylhydrazine reagent
8. Ethanol
9. Hydroxylamine hydrochloride
10. I_2-KI test solution
11. Isovaleraldehyde
12. Methanol
13. 6 M NaOH, sodium hydroxide
14. Pyridine
15. Semicarbazide reagent
16. Sodium acetate
17. Tollens' reagent (solution A and solution B)
18. Hirsch funnel
19. Hot plate
20. Neoprene adapter (no. 2)
21. Rubber stopper (no. 6, one-hole), with glass tubing
22. 50-mL side-arm filter flask
23. 250-mL side arm filter flask
24. Vacuum tubing (heavy walled)

EXPERIMENT 15

NAME _____   SECTION _____   DATE _____

PARTNER _____   GRADE _____

PRE-LAB QUESTIONS

1. Write a general structure for the following functional groups:

    a. an aldehyde

    b. a ketone

    c. a methyl ketone

    d. an aromatic aldehyde

2. What reagent gives a characteristic test for a methyl ketone?

3. Tollens' reagent will test for which functional group?

4. Write the structures of the oxime, semicarbazone and 2,4-dinitrophenylhydrazone of acetone.

EXPERIMENT 15

NAME _____ SECTION _____ DATE _____

PARTNER _____ GRADE _____

<u>REPORT SHEET</u>

Test	Isovaleraldehyde	Benzaldehyde	Cyclohexanone	Acetone	Unknown
Chromic acid					
Tollens'					
Iodoform					
2,4-Dinitro-phenyl-hydrazine					

Derivative	Observed m. p.	Literature m. p.
2,4-DNP		
Oxime		
Semicarbazone		

Unknown No._____. The unknown compound is _____.

POST-LAB QUESTIONS

1. A compound of molecular formula $C_5H_{10}O$ forms a yellow precipitate with 2,4-dinitrophenyl-hydrazine reagent and a yellow precipitate with reagents for the iodoform test. Draw a structural formula for a compound that fits these tests.

2. What kind of results do you see when the following compounds are mixed together with the given test solution?

    a.   with 2,4-dinitrophenylhydrazine

    b.   with chromic acid

    c.   with I_2-KI reagent

    d. $CH_3CH_2-\overset{\displaystyle O}{\underset{\displaystyle \|}{C}}-H$ Tollens' reagent

3. Write the structure of the compound that gives a silver mirror with Tollens' reagent and a yellow precipitate with sodium hydroxide and iodine.

4. Using the laboratory tests of this experiment, show how you could distinguish between the following compounds:

Test   cyclohexane   cyclohexanone   hexanal

5. Write an equation for each test you used in the above problem that gave a positive result.

Carbohydrates

BACKGROUND

Carbohydrates are polyhydroxy aldehydes or ketones or compounds that yield polyhydroxy aldehydes or ketones upon hydrolysis. Rice, potatoes, bread, corn, candy, and fruits are rich in carbohydrates. A carbohydrate can be classified as a monosaccharide (for example, glucose or fructose), a disaccharide (sucrose or lactose), which consists of two joined monosaccharides, or a polysaccharide (starch or cellulose), which consists of thousands of monosaccharide units linked together. Monosaccharides exist mostly as cyclic structures containing hemiacetal or hemiketal groups. These structures in solutions are in equilibrium with the corresponding open chain structures bearing aldehyde or ketone groups. Glucose, blood sugar, is an example of a polyhydroxy aldehyde (Fig. 16.1).

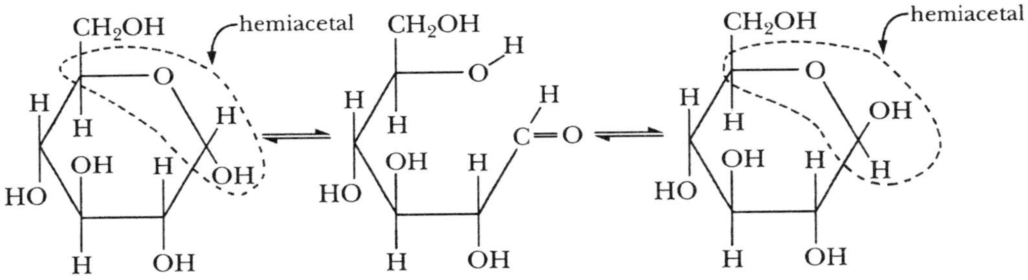

Figure 16.1 The structures of D-glucose.

Disaccharides and polysaccharides exist as cyclic structures containing functional groups such as: hydroxyl groups, acetal (or ketal), hemiacetal (or hemiketal). Most of the di-, oligo- or polysaccharides have two distinct ends. The one end which has a hemiacetal (or hemiketal) on its terminal is called the reducing end, and the one which does not contain a hemiacetal (or hemiketal) terminal is the non-reducing end. The name "reducing" is given because hemiacetals and, to a lesser extent, hemiketals can reduce an oxidizing agent such as Benedict's reagent.

The following is an example:

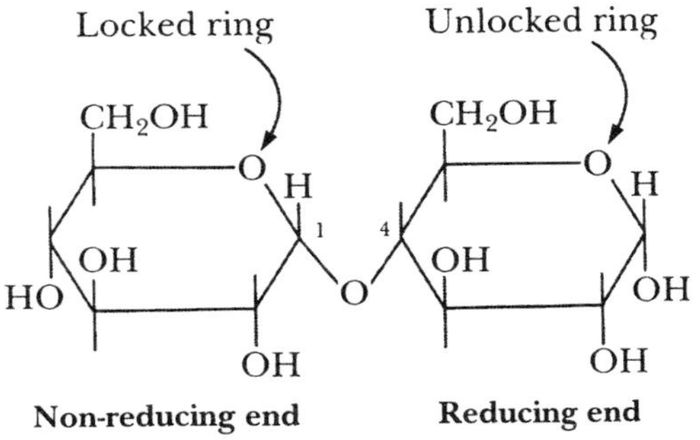

Figure 16.2 The structure of maltose, a disaccharide.

Not all disaccharides or polysaccharides contain a reducing end. An example is sucrose which does not have a hemiacetal or hemiketal group on either of its ends.

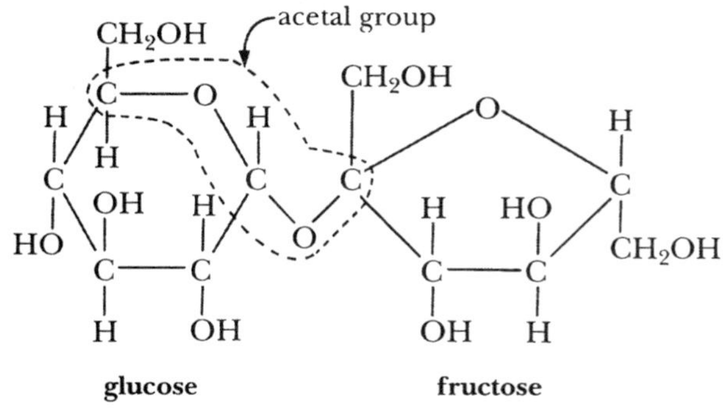

Figure 16.3 The structure of sucrose.

Polysaccharides, such as amylose or amylopectin, do have a hemiacetal group on one of their terminal ends, but practically they are non-reducing substances, because there is only one re-ducing group for every 2,000-10,000 monosaccharidic units. In such a low concentration the reducing group does not give a positive test with Benedict's or Fehling's reagent.

On the other hand, when a non-reducing disaccharide (sucrose) or a polysaccharide such as amylose is hydrolyzed, the glycosidic linkages (acetal) are broken and reducing ends are created. Hydrolyzed sucrose (a mixture of D-glucose and D-fructose) will give a positive test with Benedict's or Fehling's reagent as well as hydrolyzed amylose (a mixture of glucose and glucose containing oligosaccharides). The hydrolysis of sucrose or amylose can be achieved by using a strong acid such as HCl or with the aid of biological catalysts (enzymes).

Starch can form an intense, brilliant, dark blue or violet colored complex with iodine. The straight chain component of starch, the amylose, gives a blue color while the branched com-ponent, the amylopectin, yields a purple color. In the presence of iodine the amylose forms helixes inside of which the iodine molecule assemble as long polyiodide chains. The helix form-ing branches of amylopectin are much shorter than those of amylose. Therefore the polyiodide chains are also much shorter in the amylopectin-iodine complex than in the amylose-iodine complex. The result is a different color (purple). When starch is hydrolyzed and broken down to small carbohydrate units the iodine will not give a dark blue (or puple) color. The iodine test is used in this experiment to indicate the completion of the hydrolysis.

In this experiment you will investigate some chemical properties of carbohydrates in terms of their functional groups.

1. **Reducing and non-reducing properties of carbohydrates**
    a. *Aldoses (polyhydroxy aldehydes)*. All aldoses are reducing sugars because they contain free aldehyde functional groups. The aldehydes are oxidized by mild oxidizing agents (e.g., Benedict's or Fehling's reagent) to the corresponding carboxylates. For example,

$$R-CHO + 2Cu^{2+} \xrightarrow{\text{NaOH}} R-COO^-Na^+ + Cu_2O \downarrow$$

    (from Fehling's reagent)   **red precipitate**

    b. *Ketoses (polyhydroxy ketones)*. All ketoses are reducing sugars because they have a ketone func-tional group next to an alcohol functional group. The reactivity of this specific ketone (also called α-hydroxyketone) is attributed to its ability to form an α-hydroxyaldehyde in basic media according to the following equilibrium equations:

$$\underset{\text{ketose}}{\begin{array}{c} CH_2OH \\ | \\ C=O \\ | \\ H-C-OH \\ | \end{array}} \xrightleftharpoons{\text{base}} \underset{\text{enediol}}{\begin{array}{c} CHOH \\ || \\ C-OH \\ | \\ H-C-OH \\ | \end{array}} \xrightleftharpoons{\text{base}} \underset{\text{aldose}}{\begin{array}{c} CHO \\ | \\ H-C-OH \\ | \\ H-C-OH \\ | \end{array}}$$

c. *Hemiacetal functional group (potential aldehydes)*. Carbohydrates with hemiacetal functional groups can reduce mild oxidizing agents such as Benedict's reagent because hemiacetals can easily form aldehydes through the following equiliblium equation:

$$\underset{}{\begin{array}{c} H \quad OR' \\ \diagdown \diagup \\ C \\ \diagup \diagdown \\ R \quad OH \end{array}} \rightleftharpoons \underset{}{\begin{array}{c} H \\ \diagdown \\ C=O \\ \diagup \\ R \end{array}} + \ R'OH$$

Sucrose is, on the other hand, a nonreducing sugar because it does not contain a hemiacetal functional group. Although starch has a hemiacetal functional group at one end of its molecule, it is, however, considered as a nonreducing sugar because the effect of the hemiacetal group in a very large starch molecule becomes insignificant to give a positive Benedict's test.

2. **Hydrolysis of acetal groups**. Disaccharides and polysaccharides can be converted into monosaccharides by hydrolysis. The following is an example:

$$\underset{\substack{\text{lactose} \\ \text{(milk sugar)}}}{C_{12}H_{22}O_{11}} + H_2O \xrightarrow{\text{catalyst}} \underset{\text{glucose}}{C_6H_{12}O_6} + \underset{\text{galactose}}{C_6H_{12}O_6}$$

OBJECTIVES

1. To become familiar with the reducing or nonreducing nature of carbohydrates.
2. To experience the enzyme-catalyzed and acid-catalyzed hydrolysis of acetal groups.

PROCEDURE

Reducing or nonreducing carbohydrates

Place approximately 2 mL (30 drops) of Fehling's solution (15 drops each of solution part A and solution part B) into each of five labeled tubes. Add 10 drops of each of the following carbohydrates to the corresponding test tubes as shown in the following table.

Test tube no.	Name of carbohydrate
1	Glucose
2	Fructose
3	Sucrose
4	Lactose
5	Starch

Place the test tubes in a boiling water bath for 5 min. A 600-mL beaker containing about 200 mL of tap water and a few boiling chips is used as the bath. Record your results. Which of those carbohydrates are reducing carbohydrates?

Hydrolysis of carbohydrates

Hydrolysis of sucrose (Acid versus base catalysis)

Place 3 mL of 2% sucrose solution in each of two labeled test tubes. To the first test tube (no. 1) add 3 mL of water and three drops of dilute sulfuric acid solution. To the second test tube (no. 2) add 3 mL of water and three drops of dilute sodium hydroxide solution. Heat the test tubes in a boiling water bath for about 5 min. Cool both solutions to room temperature and then to the contents of test tube no. 1, add dilute sodium hydroxide solution (about ten drops) until red litmus paper turns blue. Test a few drops of each of the two solutions (test tubes nos. 1 and 2) with Fehling's reagent as described before. Record your results.

Hydrolysis of starch (Enzyme versus acid catalysis)

Place 2 mL of 2% starch solution in each of two labeled test tubes. To the first test tube (no. 1), add 2 mL of your own saliva. (Use a 10-mL graduated cylinder to collect your saliva.) To the second test tube (no. 2), add 2 mL of dilute sulfuric acid. Place both test tubes in a water bath that has been previously heated to 45°C. Allow the test tubes with their contents to stand in the warm water bath for 30 min. Transfer a few drops of each solution into a spot plate or two labeled microtest tubes. (Use two clean, separate medicine droppers for transferring.) To each test tube, add two drops of iodine solution. Record the color of the solutions.

Acid catalyzed hydrolysis of starch

Place 50 mL of starch solution in a 125-mL Erlenmeyer flask and add 10 mL of dilute sulfuric acid. Heat the solution in a boiling water bath for about 5 min. Using a clean medicine dropper, transfer about 3 drops of the starch solution to a spot plate or a microtest tube and then add 2 drops of iodine solution. Observe the color of the solution. If the solution gives a positive test with iodine solution (the solution should turn blue), continue heating. Transfer about 3 drops of the boiling solution at 5-min. intervals for an iodine test. *(Note: Rinse the medicine dropper very thoroughly before each test).* When the solution no longer gives a blue color with iodine solution, stop heating and record the time needed for the completion of hydrolysis.

Properties of Carboxylic Acids and Esters

BACKGROUND

Carboxylic acids are structurally like aldehydes and ketones in that they contain the carbonyl group. However, an important difference is that carboxylic acids contain a hydroxyl group attached to the carbonyl carbon.

$$-\overset{\underset{\displaystyle \|}{O}}{C}-OH$$

The carboxylic acid group

This combination gives the group its most important characteristic; it behaves as an acid.

As a family carboxylic acids are weak acids which ionize only slightly in water. As aqueous solutions typical carboxylic acids ionize to the extent of only one percent or less.

$$R-\overset{\underset{\displaystyle \|}{O}}{C}-OH + H_2O \rightleftharpoons R-\overset{\underset{\displaystyle \|}{O}}{C}-O^- + H_3O^+$$

At equilibrium most of the acid is present as unionized molecules. Dissociation constants, K_a, of carboxylic acids, where R is an alkyl group, are 10^{-5} or less. Water solubility depends to a large extent on the size of the R-group. Only a few low molecular weight acids (up to four carbons) are very soluble in water.

Although carboxylic acids are weak, they are capable of reacting with bases stronger than water. Thus, while benzoic acid shows limited water solubility, it reacts with sodium hydroxide to form the soluble salt sodium benzoate. (Sodium benzoate is a preservative in soft drinks.)

$$\langle\!\!\!\bigcirc\!\!\!\rangle\text{-COOH} + \text{NaOH} \rightarrow \langle\!\!\!\bigcirc\!\!\!\rangle\text{-COO}^-\text{Na}^+ + \text{H}_2\text{O}$$

Benzoic acid
Insoluble

Sodium benzoate
Soluble

Sodium carbonate, Na_2CO_3, and sodium bicarbonate, $NaHCO_3$, solutions can neutralize carboxylic acids, also.

The combination of a carboxylic acid and an alcohol gives an ester; water is eliminated. Ester formation is an equilibrium process, catalyzed by an acid catalyst.

$$CH_3CH_2CH_2\overset{\displaystyle O}{\overset{\|}{C}}-OH \ + \ CH_3CH_2OH \ \underset{}{\overset{H^+}{\rightleftharpoons}} \ H_2O \ + \ CH_3CH_2CH_2\overset{\displaystyle O}{\overset{\|}{C}}-OCH_2CH_3$$

Butyric acid   **Ethyl alcohol**   **Ethyl butyrate (Ester)**

Esterification ⟶

⟵ **Hydrolysis**

The reaction typically gives 60% to 70% of the maximum yield. The reaction is a reversible process. An ester reacting with water, giving the carboxylic acid and alcohol, is called *hydrolysis*; it is acid catalyzed. The base-promoted decomposition of esters yields an alcohol and a salt of the carboxylic acid; this process is called *saponification*. Saponification means "soap making", and the sodium salt of a fatty acid (e.g. sodium stearate) is a soap.

$$CH_3CH_2CH_2\overset{\displaystyle O}{\overset{\|}{C}}-OCH_2CH_3 \ + \ NaOH \ \longrightarrow \ CH_3CH_2CH_2\overset{\displaystyle O}{\overset{\|}{C}}-O^-Na^+ \ + \ CH_3CH_2OH$$

Ethyl butyrate   **Sodium butyrate**

Saponification

A distinctive difference between carboxylic acids and esters is in their characteristic odors. Carboxylic acids are noted for their sour, disagreeable odors. On the other hand, esters have sweet and pleasant odors often associated with fruits, and fruits smell the way they do because they contain esters. These compounds are used in the food industry as fragrances and flavoring agents. For example, the putrid ador of rancid butter is due to the presence of butyric acid, while the odor of pineapple is due to the presence of the ester, ethyl butyrate. Only those carboxylic acids of low molecular weight have odor at room temperature. Higher molecular weight carboxylic acids form strong hydrogen bonds, are solid and have a low vapor pressure. Thus, few molecules reach our nose. Esters, however, do not form hydrogen bonds among themselves; they are liquid at room temperature, even when the molecular weight is high. Thus, they have high vapor pressure and many molecules can reach our nose providing odor.

OBJECTIVES

1. To study the physical properties of carboxylic acids: solubility, acidity, aroma.
2. To prepare a variety of esters and note their odors.
3. To demonstrate saponification.

PROCEDURE

Carboxylic acids and their salts

Characteristics of Acetic Acid

1. Place into a clean, dry test tube (100 × 13 mm) 2 mL of water and 10 drops of glacial acetic acid. Note its odor. Of what does it remind you?

2. Take a glass rod and dip it into the solution. Using wide range indicator paper (pH 1 - 12), test the pH of the solution by touching the paper with the wet glass rod. Determine the value of the pH by comparing the color of the paper with the chart on the dispenser.

3. Now add 2 mL of 2 M NaOH to the solution. Cork the test tube and sharply tap it with your finger. Remove the cork and determine the pH of the solution as before; if not basic, continue to add more base, dropwise, until the solution is basic. Note the odor and compare the odor to the solution before the addition of base.

4. By dropwise addition of 3 M HCl, carefully reacidify the solution from 3 (above); test the solution as before with pH paper until the solution tests acid. Does the original odor return?

Characteristics of Benzoic Acid

1. Your instructor will weigh out 0.1 g of benzoic acid for sample size comparison. With your microspatula take some sample equivalent to the preweighed sample (an exact quantity is not important here). Add the solid to a test tube (100 × 13 mm) along with 2 mL of water. Is there any odor? Mix the solution by sharply tapping the test tube with your finger. How soluble is the benzoic acid?

2. Now add 1 mL of 2 M NaOH to the solution from 1 (above), cork and mix by sharply tapping the test tube with your finger. What happens to the solid benzoic acid? Is there any odor?

3. By dropwise addition of 3 M HCl, carefully reacidify the solution from 2 (above); test as before with pH paper until acidic. As the solution becomes acidic what do you observe?

Esterification

1. Into five clean, dry test tubes (100 × 13 mm) add 10 drops of liquid carboxylic acid or 0.1 g of solid carboxylic acid and 10 drops of alcohol according to the scheme in Table 17.1. Note the odor of each reactant.

TABLE 17.1 ACIDS AND ALCOHOLS

Test Tube No.	Carboxylic Acid	Alcohol
1	Formic	Isobutyl
2	Acetic	Benzyl
3	Acetic	Isopentyl
4	Acetic	Ethyl
5	Salicylic	Methyl

2. Add 5 drops of concentrated sulfuric acid to each test tube and mix the contents thoroughly by sharply tapping the test tube with your finger.

CAUTION!   *Sulfuric acid causes severe burns. Flush any spill with lots of water. Use gloves with this reagent.*

3. Place the test tubes in a warm water bath at 60°C for 15 min. Remove the test tubes from the water bath, cool and add 2 mL of water to each. Note that there is a layer on top of the water in each test tube. With a Pasteur pipet, take a few drops from this top layer and place on a watch glass. Note the odor. Match the ester from each test tube with one of the following odors: banana, peach, raspberry, nail polish remover, wintergreen.

Saponification

This part of the experiment can be done while the esterification reactions are being heated.

1. Place into a test tube (150 × 18 mm) 10 drops of methyl salicylate and 5 mL of 6 M NaOH. Heat the contents in a boiling water bath for 30 min. Record on the Report Sheet what has happened to the ester layer (1).

2. Cool the test tube to room temperature by placing it in an ice-water bath. Determine the odor of the solution and record your observation on the Report Sheet (2).

3. Carefully add 6 M HCl to the solution, 1 mL at a time, until the solution is acidic. After each addition, mix the contents and test the solution with litmus. When the solution is acidic what do you observe? What is the name of the compound formed? Answer these questions on the Report Sheet (3).

CHEMICALS AND EQUIPMENT

1. Glacial acetic acid
2. Benzoic acid
3. Formic acid
4. Salicylic acid
5. Benzyl alcohol
6. Ethanol (ethyl alcohol)
7. 2- Methyl-1-propanol (isobutyl alcohol)
8. 3-Methyl-1-butanol (isoamyl alcohol)
9. Methanol (methyl alcohol)
10. Methyl salicylate
11. 3 M HCl
12. 6 M HCl
13. 2 M NaOH
14. 6 M NaOH
15. Concentrated H_2SO_4
16. pH paper (broad range pH 1 - 12)
17. Litmus paper
18. Pasteur pipet
19. Hot plate

EXPERIMENT 17

NAME _____ SECTION _____ DATE _____

PARTNER _____ GRADE _____

PRE-LAB QUESTIONS

1. Write the structures of the following carboxylic acids:

   a. salicylic acid

   b. benzoic acid

   c. acetic acid

2. Write the products from the reaction of formic acid and sodium hydroxide.

3. n-Propyl acetate has the odor of pears. What alcohol and carboxylic acid would you use to synthesize this ester?

4. Esters can be decomposed by either hydrolysis or saponification. Explain the difference between the methods in terms of conditions and end products.

EXPERIMENT 17

NAME _____ SECTION _____ DATE _____

PARTNER _____ GRADE _____

REPORT SHEET

Carboxylic Acids and Their Salts

Characteristics of Acetic Acid

Property	Water Solution	NaOH Solution	HCl Solution
Odor			
Solubility			
pH			

Characteristics of Benzoic Acid

Property	Water Solution	NaOH Solution	HCl Solution
Odor			
Solubility			
pH			

Esterification

Test Tube	Acid	Odor	Alcohol	Odor	Ester	Odor
1	Formic		Isobutyl			
2	Acetic		Benzyl			
3	Acetic		Isopentyl			
4	Acetic		Ethyl			
5	Salicylic		Methyl			

Saponification

What has happened to the ester layer?

What has happened to the odor of the ester?

What forms on reacidification of the solution? Name the compound.

Write the chemical equation for the saponification of methyl salicylate.

POST-LAB QUESTIONS

1. Explain why acetic acid has an odor, but benzoic acid does not.

2. Write equations for each of the five esterification reactions.

   a.

   b.

   c.

   d.

   e.

3. Butyric acid has a putrid odor (like rancid butter). Suppose you got some on your hands. How could you rid your hands of the odor? (Remember, butyric acid has marginal solubility in water.)

Preparation of Isopentyl Acetate (Banana Oil)

BACKGROUND

A class of compounds called esters are often associated with the flavor and fragrance of fruits and flowers. These compounds are pleasant smelling and may be described as "fruity." Some esters, along with the corresponding flavors and fragrances, are found in Table 18.1.

The Fischer esterification is a general reaction for the preparation of esters. In this method an alcohol and a carboxylic acid, catalyzed by mineral acid, react to form an ester; water is eliminated in the process.

$$\underset{\textbf{Carboxylic acid}}{R-\overset{\overset{\displaystyle O}{\|}}{C}-OH} + \underset{\textbf{Alcohol}}{R'OH} \underset{}{\overset{H^+}{\rightleftharpoons}} \underset{\textbf{Ester}}{R-\overset{\overset{\displaystyle O}{\|}}{C}-OR'} + H_2O$$

The reaction is reversible. Equilibrium is reached rapidly through the use of the mineral acid catalyst. Since it is an equilibrium reaction, the yield of ester can be increased by shifting the equilibrium to the right, either by using an excess of one reagent or by removing one of the products, usually water.

In this experiment, the ester isopentyl acetate will be synthesized. This is a common solvent, often referred to as "banana oil", owing to its familiar odor.

$$\underset{\textbf{Acetic acid}}{CH_3-\overset{\overset{\displaystyle O}{\|}}{C}-OH} + \underset{\textbf{Isopentyl alcohol}}{CH_3-\overset{\overset{\displaystyle CH_3}{|}}{CH}-CH_2CH_2OH} \overset{H^+}{\rightleftharpoons} \underset{\textbf{Isopentyl acetate}}{CH_3-\overset{\overset{\displaystyle O}{\|}}{C}-O-CH_2CH_2-\overset{\overset{\displaystyle CH_3}{|}}{CH}-CH_3} + H_2O$$

The excess reagent in this procedure is acetic acid since it is easier to remove from the equilibrium mixture than the isopentyl alcohol; sulfuric acid is the mineral acid catalyst.

The mineral acid serves to protonate the carboxylic acid.

$$CH_3-\overset{\overset{\displaystyle :O:}{\|}}{C}-\overset{..}{O}H \rightleftharpoons \overset{H^+}{} \left[CH_3-\overset{\overset{\displaystyle :O^{+}\diagup H}{\|}}{C}-\overset{..}{O}H \longleftrightarrow CH_3-\overset{\overset{\displaystyle \overset{..}{O}\diagup H}{|}}{\underset{+}{C}}-\overset{..}{O}H \right]$$

TABLE 18.1 ESTERS ASSOCIATED WITH FLAVORS AND FRAGRANCES

Structure	Name	Flavor
$CH_3COOCH_2-\langle\bigcirc\rangle$	Benzyl acetate	Peach
$CH_3COOCH_2CH_2CH3$	n-Propyl acetate	Pear
$CH_3COOCH_2CH_2CH(CH_3)_2$	Isopentyl acetate	Banana
$CH_3COO(CH_2)_7CH_3$	Octyl acetate	Orange
$CH_3(CH_2)_2COOCH_2CH_3$	Ethyl butyrate	Pineapple
$CH_3(CH_2)_2COOCH_3$	Methyl butyrate	Apple
$HCOOCH_2CH_3$	Ethyl formate	Rum
$HCOOCH_2CH(CH_3)_2$	Isobutyl formate	Raspberry
(structure with NH_2 and $COOCH_3$ on benzene ring)	Methyl anthranilate	Grape
(structure with OH and $COOCH_3$ on benzene ring)	Methyl salicylate	Wintergreen

This increases the electrophilic character of the carbonyl carbon atom and enhances attack by the nucleophilic oxygen of the alcohol.

tetrahedral intermediate

The intermediate that results from the addition is tetrahedral and is in equilibrium with a form that has a protonated hydroxyl group; this is a potential water molecule. Elimination of a proton (which returns to solution) and the loss of water completes the reaction.

OBJECTIVES

1. To demonstrate a Fischer esterification by the preparation of isopentyl acetate.
2. To analyze the purity of the product by gas chromatography.
3. To characterize the product by infrared spectroscopy.

PROCEDURE

Reflux

1. Weigh a clean, dry 50-mL round bottom flask to the nearest 0.01 g; report the weight on your Report sheet (1). Add 10.0 mL of isopentyl alcohol (density = 0.813 g/mL; MW = 88.15); reweigh the flask, and by substraction, determine the weight of the alcohol; record on the Report Sheet to the nearest 0.01 g (2) and (3). Add 15.0 mL of acetic acid (density = 1.06 g/mL; MW = 60.05). Carefully add 1.0 mL of concentrated sulfuric acid dropwise.

CAUTION!   *When working with sulfuric acid wear gloves and dispense it in the hood. Sulfuric acid can cause severe burns should it come into contact with your skin or clothing. Should you get some on your person, remove the article of clothing and flush the area with plenty of water for 10 min. Inform your instructor and seek medical assistance.*

Swirl the solution to mix. Add two (2) boiling stones.

2. Attach the flask to a reflux assembly (Fig. 18.1) using a water-cooled condenser. On top of the condenser is a calcium chloride drying tube. The flask is heated by a heating mantle controlled by a Variac.

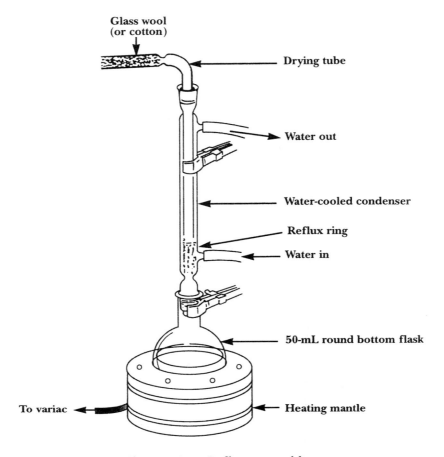

Figure 18.1 Reflux assembly.

3. Turn on the water spigot and adjust to a slow flow of water. Bring the mixture to a gentle boil so that the "reflux ring" is no more than half-way up the condenser. Continue heating at reflux for 60 min.

4. At the end of the reflux, remove the heating mantle and allow the system to cool to room temperature.

Isolation

1. Remove the reaction flask from the reflux assembly. Slowly add 15.0 mL of 5% aqueous sodium bicarbonate. Swirl the mixture to facilitate gas evolution.

2. When gas evolution slows, transfer the solution to a 125-mL separatory funnel (see Fig. 6.1, pg. 59, for technique). Shake gently and vent occasionally by opening the stopcock, until there is no longer any gas evolution. Place the separatory funnel on a ring clamp (see Fig. 6.2, pg. 59) and allow to stand for 5 min. before drawing off the lower aqueous layer. Add a fresh 15.0 mL of 5% aqueous sodium bicarbonate; shake and vent as before. Again draw off the lower layer after the layers have separated. Repeat with an additional 15.0 mL of 5% aqueous sodium bicarbonate.

3. Pour the upper organic layer out the top of the separatory funnel into a 25-mL Erlenmeyer flask. Add 0.5 g of granular, anhydrous sodium sulfate; swirl the mixture. If the drying agent clumps or the ester is cloudy, add an additional 0.5 g of anhydrous sodium sulfate, swirl and allow to stand for 10 min.

4. Transfer the ester with a Pasteur pipet to a clean, dry 25-mL round bottom flask. Be careful not to transfer any granules of sodium sulfate. Add two (2) boiling stones.

Distillation

1. Attach the 25-mL round bottom flask to a distillation assembly (see Fig. 3.2, pg. 24) for down-ward distillation. Be sure that all the glassware is clean and thoroughly dry. The receiving flask should be a clean, dry preweighed 25-mL round bottom flask (5).

2. Heat with a heating mantle controlled by a Variac.

3. Raise the temperature until the ester begins to distill. Collect the liquid distilling in the range 138 - 143°C.

4. When distillation is complete (do not distill to dryness in the distilling flask) remove the heating mantle. Disconnect the receiving flask and reweigh (6). By subtraction determine the weight of the isopentyl acetate (7). Calculate the percentage yield (9).

5. Place the sample in a vial of appropriate size. Make a neat label which contains the name of the product, the boiling range, the yield in grams, the percentage yield and your name. Turn in the sample to your instructor.

Analysis (optional)

1. Obtain an infrared spectrum of the product. Use sodium chloride plates as described in Appendix I. Compare your spectrum with the spectrum of starting material (Fig. 18.2) and that of the product (Fig. 18.3) (10). Submit your spectrum with your Report Sheet.

2. Obtain a gas chromatogram of your product (see Expt. 3 for discussion). [A Hewlett-Packard Model 5890 Chromatograph using a HP-1 crosslinked methyl silicone gum column (30 m $\times$ 0.53 mm $\times$ 2.65 μm film thickness) at a column temperature of 170°C showed isopentyl alcohol eluting at a retention time of 0.8 min. and isopentyl acetate eluting at a retention time of 1.0 min.] Record your results on the Report Sheet (11).

CHEMICALS AND EQUIPMENT

1. 5% aqueous sodium bicarbonate, $NaHCO_3$
2. Calcium chloride, anhydrous, $CaCl_2$
3. Concentrated sulfuric acid, H_2SO_4
4. Glacial acetic acid
5. Isopentyl alcohol
6. Sodium sulfate, anhydrous, granular, Na_2SO_4
7. Boiling chips
8. Heating mantle
9. Pasteur pipets
10. Ring clamp
11. 125-mL separatory funnel
12. 10- μL syringe
13. Variac
14. Vial
15. Reflux/distillation kit

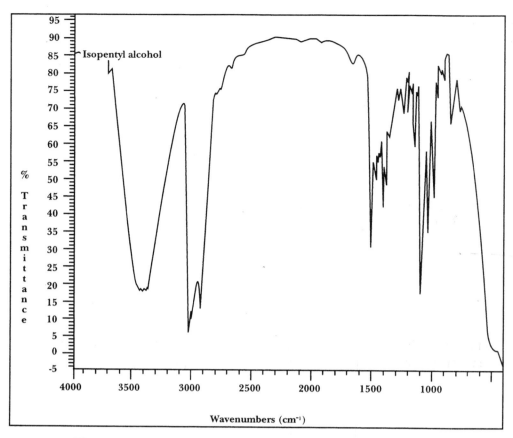

Figure 18.2 Infrared spectrum of isopentyl alcohol, neat.

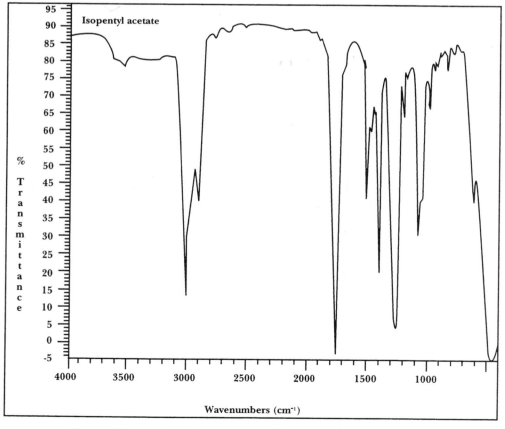

Figure 18.3 Infrared spectrum of isopentyl acetate, neat.

Polymerization Reactions

BACKGROUND

Polymers are giant molecules made of many (poly-) small units. The starting material, which is a single unit, is called the monomer. Many of the most important biological compounds are polymers. Cellulose and starch are polymers of glucose units, proteins are made of amino acids, and nucleic acids are polymers of nucleotides. Since the 1930s a large number of man-made polymers have been manufactured. They contribute to our comfort and gave rise to the previous slogan of DuPont Co.: "Better living through chemistry." Man-made fibers such as nylon and polyesters, plastics such as the packaging materials made of polyethylene and polypropylene films, polystyrene and polyvinyl chloride, just to name a few, all became household words. Man-made polymers are parts of buildings, automobiles, machinery, toys, appliances, etc.; we encounter them daily in our life.

We focus our attention in this experiment on man-made polymers and the basic mechanism by which some of them are formed. The two most important types of reactions that are employed in polymer manufacturing are the addition and condensation polymerization reactions. The first is represented by the polymerization of styrene and the second by the formation of nylon.

Styrene is a simple organic monomer which by its virtue of containing a double bond can undergo addition polymerization:

$$H_2C=CH + H_2C=CH \longrightarrow H_3C-CH-CH=CH$$

The reaction is called an addition reaction because two monomers are added to each other with the elimination of a double bond. However, the reaction as such does not go without the help of an unstable molecule, called an initiator, that starts the reaction. Benzoyl peroxide or t-butyl benzoyl peroxide are such initiators. Benzoyl peroxide splits into two halves under the influence of heat or U.V. light and thus produces two free radicals. A **free radical** is a molecular fragment that has one unpaired electron.

Thus, when the central bond was broken in the benzoyl peroxide each of the shared pair of electrons went with one half of the molecule, each containing an unpaired electron.

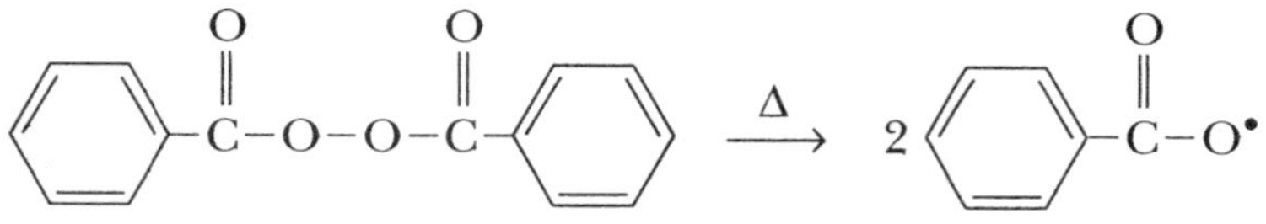

Benzoyl peroxide

Similarly, t-butyl benzoyl peroxide also gives two free radicals:

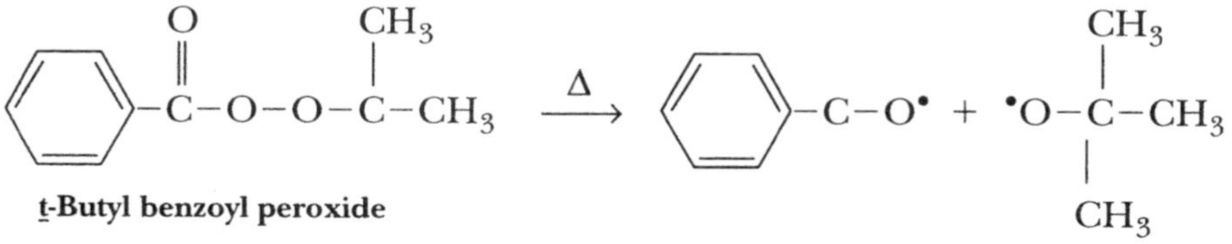

t-Butyl benzoyl peroxide

The dot indicates the unpaired electron. The free radical reacts with styrene and initiates the reaction:

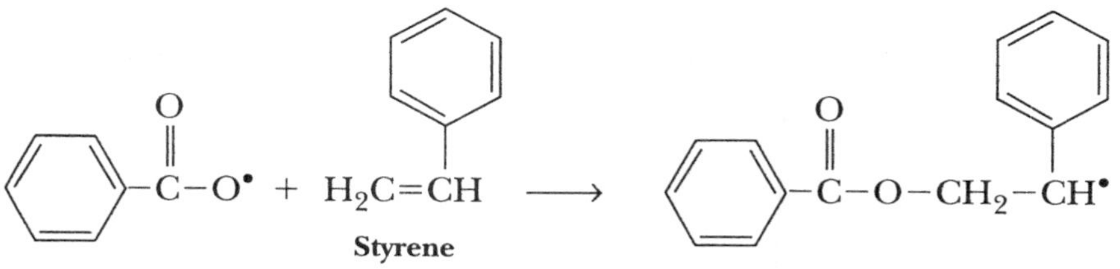

Styrene

After this, the styrene monomers are added to the growing chain one by one until giant molecules containing hundreds and thousands of styrene-repeating units are formed. Please note the distinction between the monomer and the repeating unit. The monomer is the starting material, the repeating unit is part of the polymer chain. Chemically they are not identical. In the case of styrene the monomer contains a double bond, the repeating unit (in the brackets in the following structure) does not.

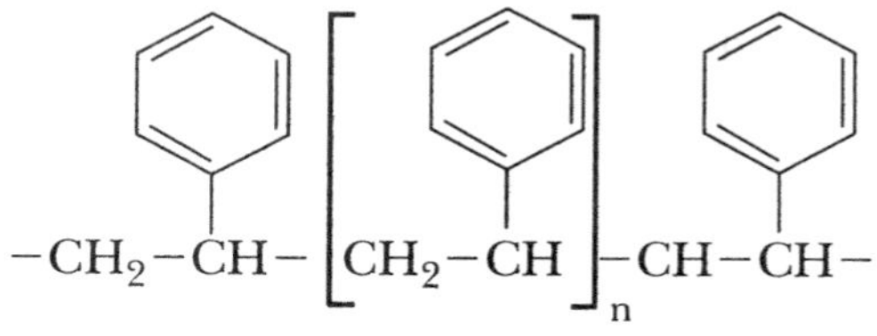

Polystyrene

Since the initiators are unstable compounds; care should be taken not to keep them near flames or heat them directly. Even dropping the bottle containing a peroxide initiator may create a minor explosion.

The second type of reaction is called a condensation reaction because we condense two monomers into a longer unit and at the same time we eliminate- expel- a small molecule. Nylon 6-6 is made of adipoyl chloride and hexamethylene diamine:

$$n \text{ Cl}-\overset{\overset{\displaystyle O}{\|}}{C}-CH_2-CH_2-CH_2-CH_2-\overset{\overset{\displaystyle O}{\|}}{C}-Cl \ + \ n \text{ H}_2N-CH_2-CH_2-CH_2-CH_2-CH_2-CH_2-NH_2 \rightarrow$$

Adipoyl chloride   **Hexamethylene diamine**

$$\text{Cl}-\overset{\overset{\displaystyle O}{\|}}{C}(CH_2)_4-\overset{\overset{\displaystyle O}{\|}}{C}-\left[NH-(CH_2)_6-NH-\overset{\overset{\displaystyle O}{\|}}{C}-(CH_2)_4-\overset{\overset{\displaystyle O}{\|}}{C}- \right]_n NH-(CH_2)_6-NH_2 \ + \ n \text{ HCl}$$

repeating unit

nylon 6-6

We form an amide linkage between the adipoyl chloride and the amine with the elimination of HCl. The polymer is called nylon 6-6 because there are six carbon atoms in the acyl chloride and six carbon atoms in the diamine. Other nylons, such as nylon 6-10, are made of sebacoyl chloride (a ten carbon containing acyl chloride) and hexamethylene diamine (six carbon atoms). We use acyl chloride rather than carboxylic acid to form the amide bond because the former is more reactive. NaOH is added to the polymerization reaction in order to neutralize the HCl that is released every time an amide bond is formed.

The length of the polymer chain formed in both reactions depends on environmental conditions. Usually the chains formed can be made longer by heating the products longer. This process is called curing.

OBJECTIVES

1. To acquaint students with the conceptual and physical distinction between monomer and polymer.
2. To perform addition and condensation polymerizations and solvent casting of films.

PROCEDURE

Preparation of Polystyrene

1. Set up your hot plate in the hood. Place 25 mL styrene in a 150-mL beaker. Add 20 drops of t-butyl benzoyl peroxide (t-butyl peroxide benzoate) initiator. Mix the solution.

2. Heat the mixture on the hot plate to about 140°C. The mixture will turn yellow.

3. When bubbles appear remove the beaker from the hot plate with beaker tongs. The polymerization reaction is exothermic and thus, it generates its own heat. Overheating would create sudden boiling. When the bubbles disappear put the beaker back on the hot plate. But every time the mixture starts boiling **you must remove the beaker**.

4. Continue the heating until the mixture in the beaker has a syrupy consistency.

5. Pour the contents of the beaker onto a clean watch glass and let it solidify. The beaker can be

cleaned from residual polystyrene by adding xylene and warming it on the hot plate under the hood until the polymer is dissolved.

6. Pour a few drops of the warm xylene solution on a microscope slide and let the solvent evaporate. A thin film of polystyrene will be obtained. This is one of the techniques—the so-called solvent-casting technique—used to make films from bulk polymers.

7. Discard the remaining xylene solution into a special jar labeled "Waste." You can wash your beaker with soap and water.

8. Investigate the consistency of the solidified polystyrene on your watch glass. You can remove the solid mass by prying it off with a spatula.

Preparation of Nylon

1. Set up a 50-mL reaction beaker and clamp above it a cylindrical paper roll (from toilet paper) or a stick.

2. Add 2.0 mL of 20% NaOH solution and 10 mL of a 5% aqueous solution of hexamethylene diamine.

3. Take 10 mL of 5% adipoyl chloride solution in cyclohexane with a pipet or syringe. Layer the cyclohexane solution slowly on top of the aqueous solution in the beaker. Two layers will form and nylon will be produced at the interface (Fig. 19.1).

4. With a bent wire first scrape off the nylon formed on the walls of the beaker.

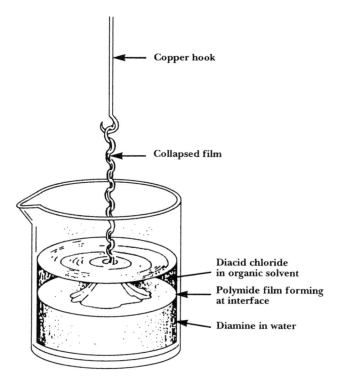

Copper hook

Collapsed film

Diacid chloride in organic solvent

Polymide film forming at interface

Diamine in water

Figure 19.1 Preparation of nylon.

5. Then slowly lift the film from the center. Pull it slowly. If you pull it too fast, the nylon rope will break.

6. Wind it around the paper roll or stick two to three times. Do not touch it with your hands.

7. Slowly rotate the roll or the stick and wind at least a 1-meter nylon rope.

8. Cut the rope and transfer the wound rope into a beaker filled with water (or 50% ethanol). Watch as the thickness of the rope collapses. Dry the rope between two filter papers.

9. There are still monomers left in the beaker. Mix the contents vigorously with a glass rod. Note the beads of nylon that have formed.

10. Pour the mixture into a cold water bath and wash it. Dry the nylon between two filter papers. Note the consistency of your products.

11. Dissolve a small amount of nylon in 80% formic acid. Place a few drops of the solution onto a microscope slide and evaporate the solvent under the hood.

12. Compare the appearance of the solvent cast nylon film with that of the polystyrene.

CHEMICALS AND EQUIPMENT

1. Styrene
2. Hexamethylene diamine solution
3. Adipoyl chloride solution
4. Sodium hydroxide solution
5. Xylene
6. Formic acid solution
7. <u>t</u>-Butyl peroxide benzoate initiator
8. Hot plate
9. Paper roll or stick
10. Bent wires
11. 10-mL pipets
12. Spectroline pipet filler
13. Beaker tongs

EXPERIMENT 19

NAME _____ **SECTION** _____ **DATE** _____

PARTNER _____ **GRADE** _____

PRE-LAB QUESTIONS

1. Why should you <u>not</u> expose <u>t</u>-butyl peroxide to direct heat?

2. Write the structure of <u>t</u>-butyl peroxide.

3. Write the reaction for the polymerization of tetrafluoroethene. Show the repeating unit of the resulting polymer (Teflon).

4. Write the structure of the monomers and that of the repeating unit in nylon 6-10.

5. Why do we call nylon a condensation polymer?

EXPERIMENT 19

NAME _____   SECTION _____   DATE _____

PARTNER _____   GRADE _____

<u>REPORT SHEET</u>

1. Describe the appearance of polystyrene and nylon.

2. Could you distinguish between polystyrene and nylon on the basis of solubility?

3. Is there any difference in the appearance of the solvent cast films of nylon and polystyrene?

POST-LAB QUESTIONS

1. A polyester is made of sebacoyl chloride and ethylene glycol,

$$\underset{\text{Cl}-\overset{\displaystyle O}{\overset{\|}{C}}(CH_2)_8\overset{\displaystyle O}{\overset{\|}{C}}-Cl}{} \quad + \quad \underset{\text{CH}_2\text{OH}}{\overset{\text{CH}_2\text{OH}}{|}} \quad \longrightarrow$$

a. Draw the structure of the polyester formed.

b. What molecules have been eliminated in this condensation reaction?

2. Two compounds , $Cl-\overset{O}{\overset{\|}{C}}$ ⟨benzene ring⟩ $\overset{O}{\overset{\|}{C}}-Cl$ in cyclohexane and $H_2N-(CH_2)_4-NH_2$ in water are reacted. Write the structure of the polyamide rope formed.

3. Since the polymerization of styrene is an exothermic reaction, why do you need to heat the mixture to 140°C?

Use of the Grignard Reagent to Prepare Benzoic Acid

BACKGROUND

The Grignard reagent is an example of an organomagnesium compound. It is prepared by reacting an alkyl or aryl halide with magnesium metal in an anhydrous ether solvent system. The reagent behaves as if its

$$R-X \ + \ Mg \ \xrightarrow{\text{anh. ether}} \ R-MgX$$

structure were of the form RMgX. This versatile reagent was discovered by Victor Grignard and earned him the Nobel prize in 1912.

The R-group can take on a variety of forms. It can be a primary (1), secondary (2) or tertiary (3) alkyl group, or an allylic (4), benzylic (5) or propargylic (6) system.

$$R = R'-CH_2-, \ R'-\overset{R''}{\underset{}{CH}}-, \ R''-\overset{R''}{\underset{R'}{C}}-, \ CH_2=CHCH_2-, \ \bigcirc\!\!\!-CH_2-, \ HC\equiv CCH_2-$$

$$(1) \qquad (2) \qquad (3) \qquad (4) \qquad (5) \qquad (6)$$

As an aryl, it can be a phenyl (7), an alkyl substituted phenyl (8) or an arene (9).

$$Ar = \qquad (7) \qquad , \qquad (8) \qquad , \qquad (9)$$

The halides used, in decreasing order of reactivity, are I > Br > Cl. The ether solvents most frequently used are diethyl ether and tetrahydrofuran.

The reagent is highly polar and has the characteristics of a carbanion; as such, the reagent may be thought of as a partially ionic species with Lewis base properties. The Lewis base character of the reagent makes it

$$\overset{\delta-}{R}\cdots\cdots\overset{\delta+}{MgX}$$

highly reactive towards molecules possessing active, acidic hydrogens. Water, alcohols, phenols, carboxylic acids, and even hydrocarbons with acidic hydrogens (for example, terminal acetylenes) will react with Grignard reagents. The reaction destroys the reagent and produces a hydrocarbon. For example,

$$\overset{\delta-}{R}-\overset{\delta+}{MgX} \quad + \quad \overset{\delta+ \,\, \delta-}{H-OH} \quad \longrightarrow \quad RH \quad + \quad HOMgX$$

Grignard **Compound with** **Hydrocarbon**
active hydrogen

$$CH_3MgI \quad + \quad H-OH \quad \longrightarrow \quad CH_4 \quad + \quad HOMgI$$

Thus, the need for anhydrous conditions in the reaction medium, and the limitation that no active hydrogens be present in any of the substrates being used is evident.

The carbanionic character of the reagent enables it to function as a nucleophile, and as such, is responsible for its use in a variety of organic syntheses. Nucleophilic attack of carbonyls is an important class of these reactions. The carbon of the carbonyl has electrophilic character and is attacked by the nucleophile. In syntheses with Grignard reagents, a two step sequence leads to a hydroxyl product.

step 1

The nucleophile attacks the electrophilic and forms a strong
of the Grignard carbon of the carbon to carbon
reagent carbonyl group bond

step 2

The intermediate is hydrolyzed by to a hydroxyl
magnesium salt dilute aqueous acid product

The Grignard reagent can be used with the carbonyl substrates listed in Table 20.1, and can lead to the products indicated (details for these reactions are found in your textbook).

TABLE 20.1 REACTIONS OF RMgx WITH CARBONYL SUBSTRATES

Carbonyl	Product
Formaldehyde ($H_2C{=}O$)	Primary alcohol (RCH_2OH)
Aldehyde ($R'CH{=}O$)	Secondary alcohol ($RR'CHOH$)
Ketone ($R'_2C{=}O$)	Tertiary alcohol (RR'_2COH)
Ester ($R'{-}\overset{\overset{\displaystyle OR''}{\mid}}{C}{=}O$)	Tertiary alcohol ($R_2R'COH$)
Carbon dioxide ($O{=}C{=}O$)	Carboxylic acid ($RCOOH$)

The reaction we will study is the reaction of the Grignard reagent, phenyl magnesium bromide, C_6H_5Br, with carbon dioxide, CO_2. The product after work-up is benzoic acid, C_6H_5COOH.

Phenyl magnesium bromide   **Carbon dioxide**   **Benzoic acid**

The Grignard reagent is prepared by reacting bromobenzene with metallic magnesium. The reaction begins slowly, but since formation of the Grignard reagent is exothermic, the ether begins to boil. The solution starts out cloudy, and as the magnesium disappears, turns gray-black. As discussed above, conditions must be water-free. Thus, the ether solvent is anhydrous and the system is protected from atmospheric water by a calcium chloride drying tube. Once formed, the phenyl magnesium bromide is used at once (*it is not stored*) and poured over solid carbon dioxide. Work-up with aqueous acid yields the product benzoic acid.

An impurity that complicates the synthesis is the heat-induced coupling of the Grignard reagent with unreacted bromobenzene to form biphenyl.

Biphenyl

This non-polar hydrocarbon is separated from the salt of the product by taking advantage of its solubility in ether.

OBJECTIVES

1. To prepare a Grignard reagent.
2. To use the Grignard reagent in the synthesis of a carboxylic acid.
3. To characterize the product by infrared spectroscopy.

PROCEDURE

Preparation of the Grignard Reagent

1. All glassware for the preparation of the Grignard reagent must be carefully dried. Dry the following glassware in an oven at 120°C for 30 min.: (1) 100-mL round bottom flask, (1) Claisen adapter, (1) 125-mL addition funnel, (1) water-cooled condenser, (2) drying tubes each containing anhydrous calcium chloride, (1) 50-mL Erlenmeyer flask. When the glassware has cooled sufficiently so you can handle each piece, lubricate the joints with silicone grease and construct the set-up shown in Fig. 20.1.

2. Obtain 1.4 g (MW = 24.31) of bright magnesium ribbon. Use forceps to handle the magnesium ribbon. Scrape both sides of the ribbon with a spatula to remove any oxide. Cut the ribbon into small pieces (approx. 5 mm). Reweigh the pieces of ribbon and record the weight to the nearest 0.01 g on the Report Sheet (1). Carefully add the pieces to the 100-mL round bottom flask. Reconnect the glassware and with a microburner, gently flame the flask. Turn off the flame and allow the glassware to cool to room temperature.

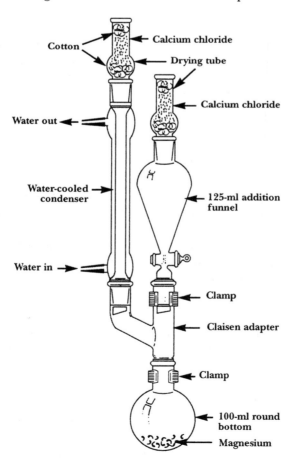

Figure 20.1 Assembly for the preparation of the Grignard reagent.

Diethyl ether (ether) is extremely flammable. Do not have any ether in the area where there is a flame. Do not begin to work with ether until all flames have been extinguished.

3. Add 15 mL of anhydrous diethyl ether to the 125-mL addition funnel; open the stopcock and drain the contents into the cooled round bottom flask.

4. Place 6.5 mL (density = 1.49 g/mL; MW = 157.01) of bromobenzene into a clean, dry pre-weighed 50-mL Erlenmeyer flask (3). Reweigh (4) and by subtraction determine the weight of the bromobenzene to the nearest 0.01 g (5); record on the Report Sheet.

5. Add 30 mL of anhydrous diethyl ether to the Erlenmeyer flask. Swirl to ensure mixing and quickly transfer to the addition funnel. Open the stopcock and add in one portion approximately one-third of the bromobenzene-ether mixture to the round bottom flask. Swirl the mixture for 2 - 3 min. using the palm of your hand to warm the mixture. Observe the mixture to determine if the reaction has started. You should see bubbles forming on the magnesium and a grey precipitate in the ether. Eventually, the exothermic reaction should cause the ether to reflux and the solution to become grey-black. Once the reaction has started, add the remaining bromobenzene-ether mixture over a 10 min. period. The rate of addition should be adjusted so that the reflux ring in the water-cooled condenser is no more than half-way up the inner tube. Keep a beaker of ice-water nearby to control the reflux should it become too vigorous. When all the solution has been added, rinse the addition funnel with 10 mL of anhydrous ether and add to the reaction mixture.

6. The reaction may not start immediately. If there is no evidence of a reaction after 5 min., there are a number of techniques you can use to initiate the reaction. Try the following procedures, one at a time, giving each a chance to work, before moving on to the next.
    a. Heat with a water-bath at 45°C until the ether boils. Remove the heat and observe whether bubbling continues without application of heat. Try this sequence 2 or 3 times.
    b. Carefully separate the Claisen adapter from the round bottom flask. With a glass stirring rod, crush 2 or 3 pieces of magnesium against the side of the reaction flask. *Do not do this so hard as to poke a hole through the glass.* Replace the Claisen adapter and heat the solution as described above.
    c. Separate the Claisen adapter from the round bottom flask and add a single, small crystal of iodine and rub the crystal gently into the metal with a glass stirring rod. Reassemble and heat as described above.
    d. To a small test tube, add 2 or 3 small pieces of magnesium metal, one mL of anhydrous ether and 5 drops of bromobenzene. With a glass stirring rod crush the magnesium metal against the glass bottom of the test tube; be careful not to poke through the bottom. The solution should bubble as the exothermic reaction begins. Transfer the entire contents of the test tube through the neck of the round bottom flask by lifting the Claisen adapter. Reassemble and swirl the contents of the flask. (Keep the joints clean during this procedure.)
    e. If after all these efforts the reaction still has not started, it may be necessary to discard and start from the beginning.

7. When the refluxing subsides and it appears that all (or most) of the magnesium has reacted, heat the mixture at reflux for 15 min. Use a heating mantle or a steam bath, but remember that diethyl ether boils at 35°C, so do not overheat. Then allow the mixture to cool. The Grignard reagent cannot be stored, so go on to the next step as soon as the solution has cooled to room temperature.

Preparation of Benzoic Acid

CAUTION!   *Dry Ice is extremely cold and will cause frostbite if it contacts skin. Use forceps, tongs, spatula or insulated gloves in handling the solid.*

1. Quickly weigh out 10 g of Dry Ice; it need not be exact. Crush by wrapping in a clean, dry towel and striking against the bench top (or hit with a hammer). Transfer the crushed Dry Ice to a 150-mL beaker. Disconnect the round bottom flask from the Claisen adapter and quickly pour the contents slowly over the crushed Dry Ice. Be aware that the reaction is vigorous and froths. Rinse the round bottom flask with 10 mL of ether and add to the beaker. Stir the mixture with a glass stirring rod until the excess has sublimed. A viscous mass should result.

2. Cool the beaker containing the ether solution in an ice-bath. Place 6 mL of concentrated hydrochloric acid in a 25-mL Erlenmeyer flask and cool in an ice-bath. Carefully add the cold acid to the cold ether solution. Stir. If all of the solid has not dissolved add more acid until everything has gone into solution. Add 10 mL of water. You should see two layers. Transfer the mixture to a 125-mL separatory funnel (see Figs. 6.1 and 6.2, pg. 59 for techniques).

3. Test the lower aqueous layer with pH paper; it should test a pH 2 - 3; if not, add more acid until the layer is acidic. Draw off the lower aqueous layer and discard. (The ether layer remains in the separatory funnel and contains the benzoic acid and the biphenyl impurity.)

4. Add to the separatory funnel a volume of water equal to the volume of ether (40 - 50 mL). Shake (be sure to vent properly). Allow the layers to separate; draw off the lower aqueous layer and discard.

5. Add 30 mL of 5% aqueous sodium hydroxide to the ether layer in the separatory funnel. Shake (be sure to vent). Allow the layers to separate. Draw off and save the lower aqueous base solution in a 250-mL beaker. Repeat the extraction an additional two times. After each extraction add the lower basic layer to the beaker. Test the sodium benzoate solution with litmus; it should test basic.

6. Pour the ether layer into a 150-mL beaker. Add 5 g of anhydrous sodium sulfate and allow to stand for 15 min. Decant the ether into a clean 150-mL beaker and evaporate the ether on a steam bath in the hood. Collect the solid biphenyl and take its melting point. Record on the Report Sheet (7).

7. Heat the aqueous base solution to 80°C on a hot plate. Stir with a glass stirring rod at that temperature for 10 min. Cool the solution in an ice-bath and then precipitate the benzoic acid by adding 9 M hydrochloric acid until a pH of 2 - 3 is obtained. The solid benzoic acid is isolated by vacuum filtration with a small Büchner funnel (see Fig. 15.1, pg. 156). Wash the solid with water and allow it to air dry on the Büchner funnel. Record on the Report Sheet the weight of the crude benzoic acid to the nearest 0.01 g (8) and the melting point (9).

8. Benzoic acid can be purified by recrystallizing from water. The solubility in water is 6.8 g/100 mL at 95°C and 0.17 g/100 mL at 0°C. Dissolve your solid in hot water in a ratio equivalent to the solubility given above. Once dissolved, let the solution cool slowly to room temperature; then complete the crystallization by placing the solution in an ice-bath. Crystals should appear as the solution cools. If crystals do not appear immediately, induce crystallization by scratching the inside glass of the container (beaker or Erlenmeyer flask) with a glass stirring

rod. Collect the product in a small Büchner funnel by vacuum filtration. Wash the solid in the Büchner funnel with 10 mL of ice-cold water. Allow the solid to air dry on the Buchner funnel. Once dry, record on the Report Sheet the weight of the crystals to the nearest 0.01 g (10) and the melting point (12).

9. Calculate the percentage yield (13).

Analysis (optional)

1. Obtain an infrared spectrum of your benzoic acid. Compare your sample to the spectrum of known material (Fig. 20.2) (14). The spectrum is obtained as a KBr mull (see Appendix I for discussion).

2. Submit your sample to the instructor in a labeled vial along with your spectrum. The label should contain your name, the name of the compound, the melting point and the percentage yield.

CHEMICALS AND EQUIPMENT

1. Bromobenzene
2. Calcium chloride, $CaCl_2$
3. Diethyl ether, anhydrous
4. Dry ice, CO_2
5. Concentrated hydrochloric acid, HCl
6. 9 M hydrochloric acid, 9 M HCl
7. Iodine, crystals, I_2
8. Magnesium, Mg, ribbon
9. 5% aqueous sodium hydroxide, NaOH
10. Sodium sulfate, anh., Na_2SO_4
11. Büchner funnel
12. Heating mantle
13. Hot plate
14. pH paper, 2 - 5 range
15. 250-mL side-arm filter flask
16. 125-mL separatory funnel
17. Silicone grease
18. Steam bath, copper
19. Variac
20. Vial
21. Grignard reaction kit

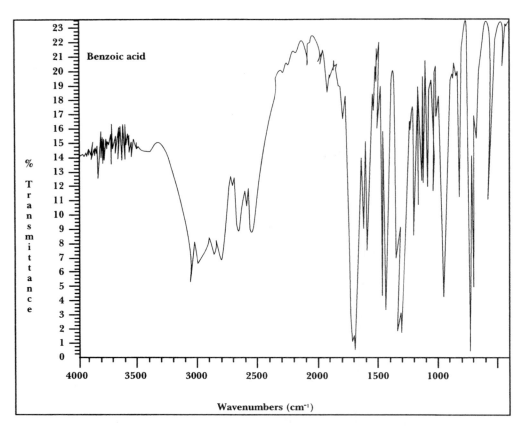

Figure 20.2 Infrared spectrum of benzoic acid, KBr.

EXPERIMENT 20

NAME _____ SECTION _____ DATE _____

PARTNER _____ GRADE _____

PRE-LAB QUESTIONS

1. In the preparation of the Grignard reagent, anhydrous conditions must be kept during the reaction. What happens if water gets into the system?

2. Benzoic acid is extracted from the diethyl ether layer by washing with sodium hydroxide. Write an equation to show how this extraction is possible. Why is the biphenyl not extracted?

3. Why is carbon dioxide a good substrate for Grignard reagents to act on?

4. Why is water an effective solvent from which to recrystallize the benzoic acid?

EXPERIMENT 20

NAME _____ SECTION _____ DATE _____

PARTNER _____ GRADE _____

<u>REPORT SHEET</u>

1. Weight of magnesium ribbon   _____ g

2. Moles of magnesium [(1)/24.31)]   _____ moles

3. Weight of 50-mL Erlenmeyer flask   _____ g

4. Weight of 50-mL Erlenmeyer flask and bromobenzene   _____ g

5. Weight of bromobenzene (4) - (3)   _____ g

6. Moles of bromobenzene [(5)/157.01)]   _____ moles

7. Melting point of biphenyl (lit. m.p. = 69 - 71°C)   _____ °C

8. Weight of crude benzoic acid   _____ g

9. Melting point of crude benzoic acid   _____ °C

10. Weight of pure benzoic acid   _____ g

11. Moles of benzoic acid [(10)/122.12)]   _____ moles

12. Melting point of pure benzoic acid (lit. m. p. = 120 - 122 °C) _____ °C

13. % Yield = $\dfrac{\text{moles of benzoic acid (11)}}{\text{moles of limiting reagent}} \times 100 =$ _____ %

14. Comparison of infrared spectra

POST-LAB QUESTIONS

1. It is not a good practice to wash glassware with acetone prior to carrying out this procedure. Why should the acetone wash be avoided? (Write an equation as part of the explanation.)

2. One method used to initiate formation of the Grignard reagent requires crushing some magnesium with a glass stirring rod. What is the rational for doing this?

3. A student attempted a reaction between equimolar quantities of phenylmagnesium bromide and p-hydroxy benzaldehyde:

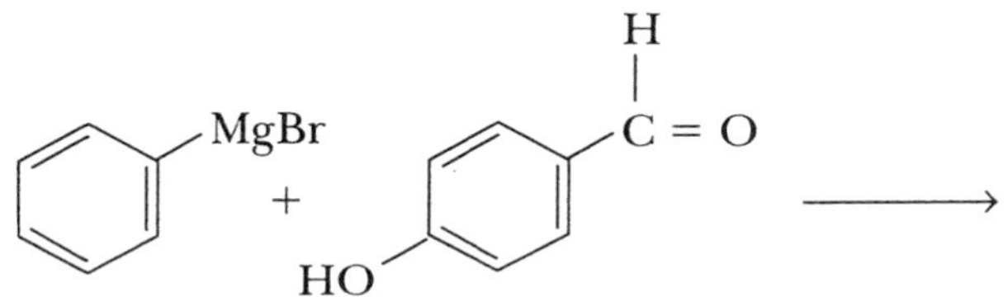

    Upon work-up, no product resulting from addition of the Grignard reagent could be found. Only unreacted starting material and benzene were isolated. Why didn't the student get a Grignard product and how did the benzene form?

4. How much Dry Ice (CO_2, in moles) was used in this reaction? Is this the limiting reagent? If not, what is the limiting reagent?

EXPERIMENT

21

Preparation of Acetyl Salicylic Acid (Aspirin)

BACKGROUND

One of the most widely used over-the-counter non-prescription drugs is aspirin. In the United States more than 15,000 pounds is sold each year. It is no wonder there is such wide use when one considers the medicinal applications for aspirin. It is an effective analgesic (pain killer) that can reduce the mild pain of headache, toothache, neuralgia (nerve pain), muscle pain and joint pain (from arthritis and rheumatism). Aspirin behaves as an antipyretic drug (it reduces fever), and an anti-inflammatory agent capable of reducing the swelling and redness associated with inflammation. It is an effective agent in preventing strokes and heart attacks due to its ability to act as an anti-coagulant.

Early studies showed the active agent that gave these properties was salicylic acid. However, salicylic acid contains the phenolic group and the carboxylic acid. As a result, the compound was too harsh to the linings of the mouth, esophagus and stomach. Contact with the stomach lining caused some hemorrhaging. The Bayer Company in Germany patented the ester acetylsalicylic acid and marketed the product as "aspirin" in 1899. Their studies showed that this material was less of an irritant; the acetylsalicylic acid was found to hydrolyze in the small intestine to salicylic acid which then was absorbed into the bloodstream. The relationship between salicylic acid and aspirin are shown in the following formulas:

Salicylic acid

Acetylsalicylic acid (Aspirin)

Aspirin still has its side effects. Hemorrhaging of the stomach walls can occur even with normal dosages. These side effects can be reduced through coatings or through the use of buffering agents. Magnesium hydroxide, magnesium carbonate, and aluminum glycinate, when mixed into the formulation of the aspirin (e.g. Bufferin), reduce the irritation.

This experiment will acquaint you with a simple synthetic problem in the preparation of aspirin. The preparative method uses acetic anhydride and an acid catalyst, like sulfuric or phosphoric acid, to speed up the reaction with salicylic acid.

| Salicylic acid | Acetic anhydride | | Aspirin | Acetic acid |

If any salicylic acid remains unreacted, its presence can be detected with a 1% ferric chloride solution. Salicylic acid has a phenol group in the molecule. The ferric chloride gives a violet color with any molecule possessing a phenol group (see Experiment 9). Notice the aspirin no longer has the phenol group. Thus, a pure sample of aspirin will not give a purple color with 1% ferric chloride.

OBJECTIVES

1. To illustrate the synthesis of the drug, aspirin.
2. To use a chemical test to determine the purity of the preparation.

PROCEDURE

Preparation of aspirin

1. Prepare a bath using a 400-mL beaker filled about half-way with water. Heat to boiling.

2. Take 2.0 g of salicylic acid and place it in a 125-mL Erlenmeyer flask. Use this quantity of salicylic acid to calculate the theoretical or expected yield of aspirin (1). Carefully add 3 mL of acetic anhydride to the flask, and then while swirling, add 3 drops of concentrated sulfuric acid.

CAUTION!   *Acetic anhydride will irritate your eyes. Sulfuric acid will cause burns to the skin. Handle both chemicals with care. Dispense in the hood. Use gloves with these reagents.*

3. Mix the reagents and then place the flask in the boiling water bath; heat for 15 min. (Fig. 21.1). The solid will completely dissolve. Swirl the solution occasionally.

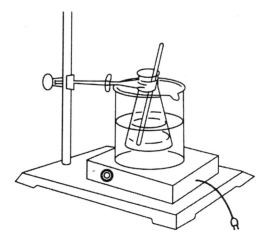

Figure 21.1 Assembly for the synthesis of aspirin.

4. Remove the Erlenmeyer flask from the bath and let it cool to approximately room temperature. Then slowly pour the solution into a 150-mL beaker containing 20 mL of ice water, mix thoroughly, and place the beaker in an ice bath. The water destroys any unreacted acetic anhydride and will cause the insoluble aspirin to precipitate from solution.

5. Collect the crystals by filtering under suction with a Büchner funnel. The assembly is shown in Fig. 21.2. (Also see fig. 15.1, pg. 156.)

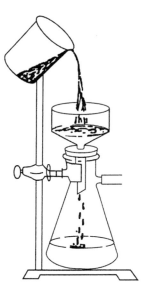

Figure 21.2 Filtering using the Büchner funnel.

6. Obtain a 250-mL filter flask and connect the side-arm of the filter flask to a water aspirator with heavy wall vacuum rubber tubing. (The thick walls of the tubing will not collapse when the water is turned on and the pressure is reduced.)

7. The Büchner funnel is inserted into the filter flask through either a filtervac, a neoprene adapter or a one-hole rubber stopper, whichever is available. Filter paper is then placed into the Büchner funnel. Be sure that the paper *lies flat* and *covers all the holes*. Wet the filter paper with water.

8. Turn on the water aspirator to maximum water flow. Pour the solution into the Büchner funnel.

9. Wash the crystals with two 5 mL portions of cold water followed by one 10 mL portion of cold ethanol.

10. Continue the suction of air through the crystals for several minutes to help dry them. Disconnect the rubber tubing from the filter flask before turning off the water aspirator.

11. Using a spatula place the crystals between several sheets of paper toweling or filter paper and press-dry the solid.

12. Weigh a 50-mL beaker (2). Add the crystals and reweigh (3). Calculate the weight of crude aspirin (4). Determine the percent yield (5).

Determine the purity of the aspirin

1. The aspirin you prepared is **not** pure enough for use as a drug and is not suitable for ingestion. The purity of the sample will be tested with 1% ferric chloride solution and compared with a commercial aspirin and salicylic acid.

2. Label three test tubes (100 x 13 mm) 1, 2, and 3; place a few crystals of salicylic acid into test tube no. 1, a small sample of your aspirin into test tube no. 2, and a small sample of a crushed commercial aspirin into test tube no. 3. Add 5 mL of distilled water to each test tube and shake to dissolve the crystals.

3. Add 10 drops of 1% aqueous ferric chloride to each test tube.

4. Compare and record your observations. The formation of a purple color indicates the presence of salicylic acid. The intensity of the color qualitatively tells how much salicylic acid is present.

CHEMICALS AND EQUIPMENT

1. Acetic anhydride
2. Concentrated sulfuric acid, H_2SO_4
3. Commercial aspirin tablets
4. 95% ethanol
5. 1% ferric chloride
6. Salicylic acid
7. Boiling chips
8. Büchner funnel, small
9. 250-mL filter flask
10. Filter paper
11. Filtervac or neoprene adaptor
12. Hot plate

Synthesis of Sulfanilamide: a Multistep Synthesis

EXPERIMENT 22

BACKGROUND

In the 1930s it was discovered that the compound sulfanilamide (1) could kill many types of harmful bacteria and thus, cure several diseases. Because sulfanilamide itself has some toxicity to humans, a number of derivatives of this compound were prepared. Sulfathiazole and sulfapyridine, among them, act the same way as sulfanilamide but have less toxicity to humans. Collectively these drugs are called "sulfa drugs." They are prescribed frequently to treat infections that are resistant to antibiotics. Sulfa drugs are effective against bacteria which synthesize folic acid from a precursor, *para*-aminobenzoic acid. Folic acid is essential in a number of biosynthetic processes, such as the synthesis of amino acids and nucleotides. The sulfa drugs resemble *para*-aminobenzoic acid (2). When the bacteria ingest sulfa drugs they mistake them for *para*-aminobenzoic acid and thus, the drug acts as a competitive inhibitor preventing the synthesis of folic acid. As a consequence the bacteria die. Since humans do not synthesize folic acid but obtain it from the diet (it is one member of the vitamin B family) sulfa drugs do not inhibit metabolic processes in humans. Sulfa drugs are potent antibacterial agents and prescribed in the form of oral or topical medication.

$$H_2N - \underset{\text{(1) Sulfanilamide}}{\bigcirc} - SO_2NH_2 \qquad H_2N - \underset{\text{(2) } para\text{-Aminobenzoic acid}}{\bigcirc} - COOH$$

Sulfa drugs are easily synthesized in a number of steps from simple raw material such as benzene (3) by classical organic methods. Such a process is called total synthesis. In the present experiment we shall synthesize sulfanilamide starting with acetanilide (6). Acetanilide itself could be obtained from benzene in four steps. We omit these four steps to save time and to avoid the use of benzene which is carcinogenic.

$$\underset{(3)}{\bigcirc} \xrightarrow[\text{H}_2\text{SO}_4]{\text{HNO}_2} \underset{(4)}{\overset{NO_2}{\bigcirc}} \xrightarrow[\text{HCl}]{\text{Fe}} \underset{(5)}{\overset{NH_2}{\bigcirc}} \xrightarrow[\text{CH}_3\text{COO}^-\text{Na}^+]{(\text{CH}_3\text{CO}_2)_2\text{O}} \underset{(6)}{\overset{\overset{\text{O}}{\underset{\|}{NHC-CH_3}}}{\bigcirc}}$$

The synthesis of sulfanilamide from acetanilide starts with the introduction of a sulfonyl chloride group onto the benzene ring. The reagent is chlorosulfonic acid, $ClSO_3H$ (7), and the reaction is known as chlorosulfonation:

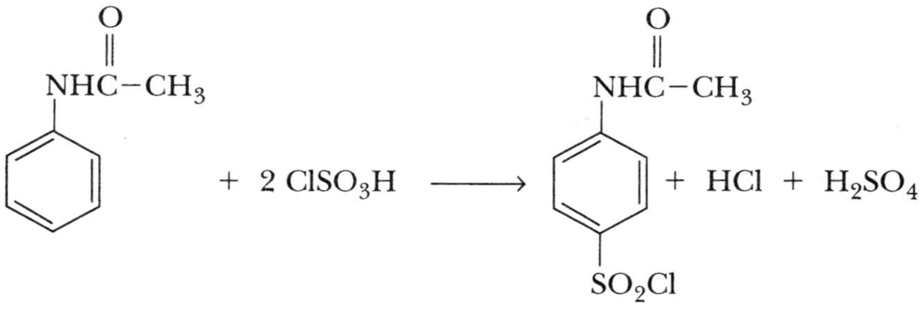

(6)   (7)   (8)

This is a typical electrophilic aromatic substitution reaction in which the acetamido group,

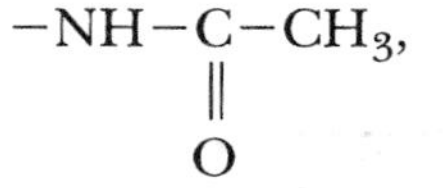

$$-NH-\underset{\underset{O}{\|}}{C}-CH_3,$$

directs the incoming group predominantly into the para position. The product, p-acetamidobenzene-sulfonyl chloride (8), is precipitated when the reaction mixture is poured into ice water and it can be isolated by filtration. The ice water also hydrolyses the excess reactant chlorosulfonic acid:

$$ClSO_3H + H_2O \longrightarrow HCl + H_2SO_4$$

In the second step of the reaction the isolated wet solid sulfonyl chloride product (8) is first converted to the corresponding sulfonamide (9) with the aid of excess ammonia. In the third step, after isolating the p-acetamidobenzenesulfonamide (9) by filtration, this compound is converted to the corresponding ammonium chloride salt (10) by acid hydrolysis:

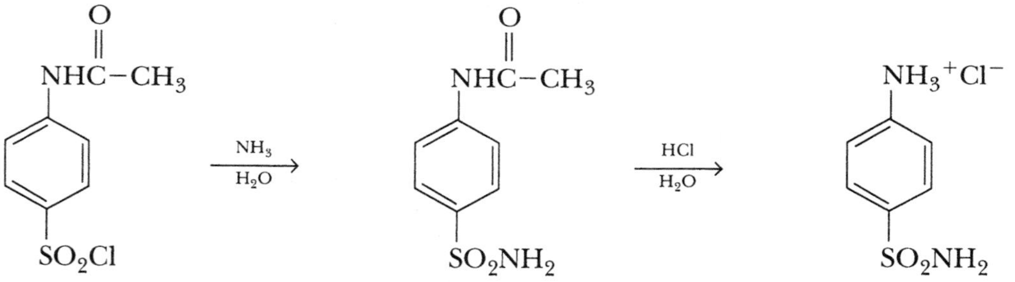

(8)   (9)   (10)

In the final step of the synthesis the ammonium salt (10) is converted to the amine by bringing the pH to 8.0 with the aid of sodium carbonate. The final product, sulfanilamide (1), can be purified by crystallization.

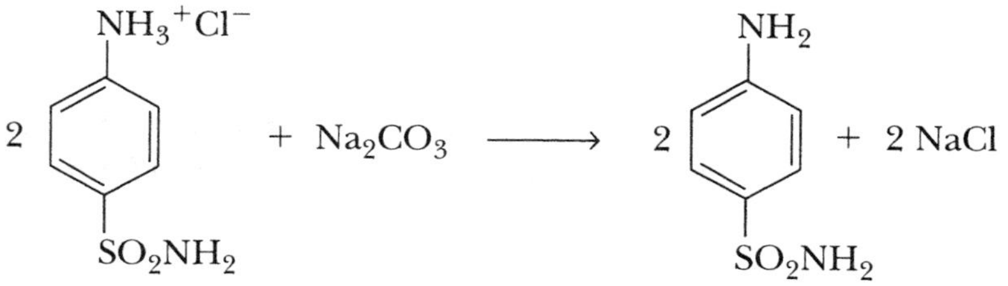

(10)   (1)

This four step organic synthesis gives a useful introduction to the procedures in which complex organic compounds are manufactured from simple raw materials.

OBJECTIVES

1. To demonstrate the synthesis of a sulfa drug.
2. To learn isolation and purification techniques.

PROCEDURE

1. Weigh approximately 1.5 g acetanilide. Record the weight to the nearest 0.01 g on your Report Sheet (1) and calculate the number of moles and record it (2). In the hood set up a clean, dry 100-mL round bottom flask with a Claisen adapter (Fig. 22.1).

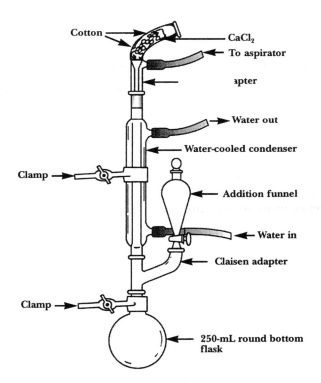

Figure 22.1 Reaction vessel with Claisen adapter.

The side arm of the Claisen adapter should be fitted with a water-cooled condenser topped with a vacuum adapter tube filled with anhydrous $CaCl_2$ (4 - 6 mesh). Connect the vacuum adapter to an aspirator by vacuum rubber tubing. Be sure to grease all the joints to provide a good seal. Add the acetanilide to the flask. Place the flask in a water bath and cool to about 12°C.

2. Take a clean and dry 125-mL separatory funnel. Make certain that the stopcock is firmly sealed. In the hood, add 11 mL chlorosulfonic acid to the separatory funnel.

CAUTION! *Chlorosulfonic acid reacts rapidly with water. Be very careful to avoid contact with your skin and, in general, with moisture. In case chlorosulfonic acid makes contact with your skin wash it immediately with copious amount of cold water and rinse it with dilute sodium bicarbonate solution. Wash all equipment carefully that contained or was in contact with chlorosulfonic acid.*

Place the stopper in the separatory funnel. Position the separatory funnel in the straight arm of the Claisen adapter so that the chlorosulfonic acid will drop directly on the acetanilide in the flask.

3. Turn on the aspirator; air will flow through the vacuum adapter. Open the stopcock of the separatory funnel completely so that the chlorosulfonic acid adds rapidly to the flask. In order to facilitate the rapid flow out of the separatory funnel you should lift slightly the stopper of the funnel. After the addition is completed replace the stopper. Swirl the reaction flask slightly. Maintain the water bath temperature below 20°C by occasionally adding pieces of ice. After all the solid has dissolved, allow the flask to warm to room temperature. Finally place the reaction vessel on a steam bath for 10 - 20 min. Occasional swirling will speed up the reaction. The reaction, is completed when upon swirling no increase in the gas evolution can be detected.

4. Using an ice water bath, cool the reaction mixture to room temperature. Fill a 250-mL beaker 3/4 full with crushed ice and add about 50 mL water. In the hood, with constant stirring, slowly pour the reaction mixture onto the ice.

CAUTION! *Avoid splattering the chlorosulfonic acid.*

Rinse the reaction vessel with a small amount of water and add it to the beaker. The product is now in the form of a precipitate in your beaker. Allow the residual ice to melt and disperse the clumps formed by stirring. Place a Büchner funnel on a side-arm filter flask and position a filter paper so that all the holes are covered. Connect the side arm of the filter flask to an aspirator and turn the water on. Flatten the filter paper on the Büchner funnel by adding a few drops of water. Filter the contents of the beaker and wash it with a few mL of cold water.

5. In the hood transfer the p-acetamidobenzene sulfonyl chloride into a 125-mL Erlenmeyer flask. Add, in small portions, with constant stirring 10 mL of concentrated (28%) aqueous ammonia. The reaction proceeds with a rapid evolution of heat. Disperse the product to form a thick slurry. Place the Erlenmeyer flask on a steam bath and heat for 30 min. To avoid ammonia fumes seeping into the lab, place an inverted funnel over the Erlenmeyer flask and connect it with rubber tubing to the aspirator (Fig. 22.2). Turn on the aspirator and keep it on to remove the ammonia vapors. After 30 min. cool the reaction mixture in an ice-water bath and add dropwise 6 M H_2SO_4 until Congo red test paper turns blue (pH 3). Cool the reaction mixture in ice-water bath and vacuum filter it on a Büchner filter as described in step 4. Wash the crystals with 2 - 3 mL cold water and dry. Scrape the crystals onto a preweighed watchglass or a plastic weighing dish. Record the tare (3) and the weight of the dish and the product (4) and the net weight (5) on your Report Sheet. Calculate the number of moles of the product and record it on your Report Sheet (6). Calculate the percentage yield of the p-acetamidobenzenesulfonamide and report it on your Report Sheet (7). Transfer a few crystals into each of two capillary tubes for melting points determinations later.

6. Transfer the products obtained above into a 100-mL roundbottom flask. Add an amount of 5 M HCl that is equal to twice the weight recorded (4) on your Report Sheet. Attach a water-cooled condenser to the flask and and connect it to an inverted-funnel gas trap (Fig. 22.3). Heat the flask on a steam bath for 30 min. allowing only a gentle reflux.

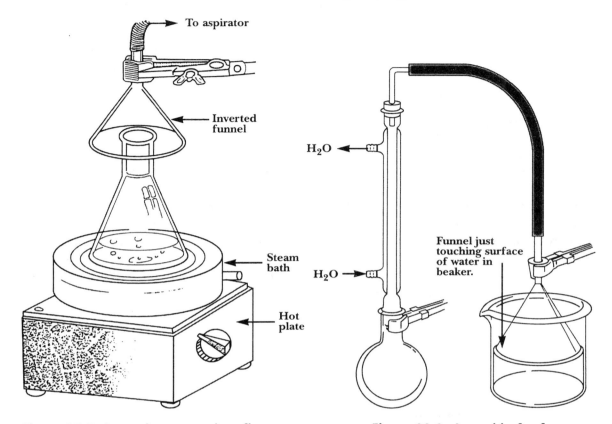

Figure 22.2 Removing ammonia reflux.   **Figure 22.3 Assembly for fumes.**

At the beginning swirl the reaction flask a few times to disperse the organic compound. At the end of 30 min. add about an equal volume of water and transfer the mixture into a 150-mL beaker. Neutralize the solution and bring it to pH 8 (blue color of litmus paper) by adding small portions of Na_2CO_3. Cool the mixture in an ice-water bath. Filter the sulfonamide crystals with a Büchner funnel under vacuum as in step 4 above. Transfer the dried crystals into a preweighed weighing dish and record the tare (8) and the weight of the tare and product (9) on your Report Sheet. Determine the net weight by subtraction (10) and calculate the moles of sulfanilamide (11). Calculate the percentage yield of your sulfanilamide product and record it on your Report Sheet (12). Take a few crystals of your sulfanilamide and insert it into two capillary tubes.

7. With the aid of a file move the samples in the capillary tubes to the closed end. Obtain the melting points, twice for each sample. Use the methods and techniques given in Experiment 4. Record on your Report Sheet the melting point of the p-acetamidobenzenesulfonamide (13) and the sulfanilamide (14).

CHEMICALS AND EQUIPMENT

1. Acetanilide
2. Chlorosulfonic acid
3. Aqueous ammonia (28 %)
4. H_2SO_4 solution (6 M)
5. Congo red test paper
6. HCl solution (5 M)
7. Na_2CO_3
8. 100-mL round-bottom flask
9. Claisen adapter
10. 125-mL separatory funnel
11. 250-mL beaker
12. Büchner funnel
13. Filter paper
14. Water-cooled condenser
15. Steam bath
16. Capillary tubes
17. Melting point apparatus

EXPERIMENT 22

NAME _____ SECTION _____ DATE _____

PARTNER _____ GRADE _____

PRE-LAB QUESTIONS

1. Name the intermediate compounds (4) and (5) in the synthesis of acetanilide.

2. What should you do if a drop of chlorosulfonic acid makes contact with your skin?

3. In the chlorosulfonation reaction, besides the main product (8), a small amount of *ortho*-acetamidobenzenesulfonyl chloride is formed. Write the structure of this compound.

4. The reaction in which compound (9) is converted to compound (10) is called hydrolysis. Write a balanced equation for this reaction.

EXPERIMENT 22

NAME _____ SECTION _____ DATE _____

PARTNER _____ GRADE _____

<u>REPORT SHEET</u>

1. Weight of acetanilide   _____ g

2. Moles of acetanilide [(1)/135.17]   _____ moles

3. Weight of weighing dish (tare)   _____ g

4. Weight of tare and product   _____ g

5. Weight of product [(9)-(8)]   _____ g

6. Moles of p-acetamidobenzenesulfonamide [(5)/214.24]   _____ moles

7. % yield of product: [(6)/(2)] × 100   _____ %

8. Weight of weighing dish (tare)   _____ g

9. Weight of tare and sulfanilamide   _____ g

10. Weight of sulfanilamide [(4)-(3)]   _____ g

11. Moles of sulfanilamide [(10)/172.20]   _____ moles

12. % yield of sulfanilamide: [(11)/(2)] × 100   _____ %

13. Melting point of p-acetamidobenzenesulfonamide   _____ °C

14. Melting point of sulfanilamide   _____ °C

POST-LAB QUESTIONS

1. Why is chlorosulfonic acid a dangerous reagent when it comes in contact with water?

2. Calculate the percentage yield for the conversion of the sulfonamide to the sulfanilamide.

3. Assume that in the process of converting the ammonium salt to the amine (last part of the Procedure step 6) you brought the pH only to 6 instead of 8. How would that effect your yield?

4. The melting point of sulfanilamide is 163°C. Lowering the melting point is caused by impurities. On the basis of your melting point obtained experimentally on sulfanilamide, what can you say regarding the purity of your preparation?

Preparation of Methyl Orange

EXPERIMENT

23

BACKGROUND

Many organic compounds are colored because they absorb light in the visible region of the electromagnetic spectrum. The visible region lies between the ultraviolet and the infrared range of the spectrum (Fig. 23.1).

Ultraviolet	Visible	Infrared
200 nm	400 nm Blue	800 nm Red

2.5 μm   15 μm

(1 nm = 10^{-9} m; 1 μm = 10^{-6} m)

Figure 23.1 A portion of the electromagnetic spectrum showing where the visible range lies.

The color we perceive is the complement of the color absorbed by the organic compound. Light coming from the sun or a lamp consists of all the wavelengths of light in the visible region, so-called "white light." When this light strikes the compound, light of a certain wavelength is absorbed by the substance and the rest are transmitted or reflected to our eyes. Thus, what is perceived (seen) is the "white light" minus the wavelengths absorbed by the organic compound. For example, a compound absorbing in the blue range (400 - 500 nm) will be seen as a yellow-orange color (white light with the blue removed).

The structural characteristics of colored organic compounds is that they have an extended conjugated system of π-electrons. There is a rough correlation between the length of the conjugated (delocalized) π-electron system and the wavelength of absorption: the longer the length of the conjugated system, the longer the wavelength of absorption. The highly conjugated system of β-carotene (1) is yellow-orange because it absorbs in the blue region with a λ_{max} at 455 nm. Pentacene (2) is a deep violet because it absorbs in the yellow-green region with a λ_{max} at 575 nm.

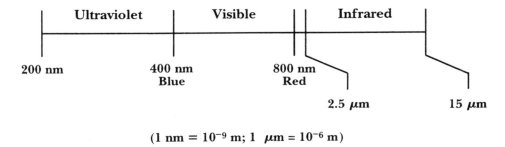

(1) β-**Carotene**   (2) **Pentacene**

(3) Congo red

(4) Para red

Azo aromatic derivatives, Ar-N=N-Ar', possess an extended conjugated system of π-electrons that absorb in the visible region of the electromagnetic spectrum. These brightly colored compounds are widely used as dyes and are generally referred to as **azo dyes**. The starting material for these dyes are aromatic amines, Ar-NH_2, and phenols, Ar-OH. Depending on the aromatic rings and the substituent groups on the rings, dyes with a broad range of colors can be synthesized. Examples of azo dyes are congo red (3), para red (4) and methyl orange (5), the compound to be synthesized in this experiment.

REACTION

Methyl orange is prepared by the diazonium coupling reaction. These reactions are typical electrophilic aromatic substitutions: the positively charged diazonium ion, Ar-N_2^+, is the electrophile that reacts with the electron-rich aromatic ring of an aryl amine (Ar-NH_2) or phenol (Ar-OH).

Sulfanilic acid (6) is a primary aromatic amine (Ar-NH_2). When treated with nitrous acid, it is converted into the diazonium cation (7). The diazotizing agent, nitrous acid, HNO_2, (or the anhydride dinitrogen trioxide, N_2O_3, in equilibrium with nitrous acid)

$$NaNO_2 + HCl \longrightarrow HONO + NaCl$$

Sodium nitrite   **Nitrous acid**

$$2\ HONO \xrightarrow{\longleftarrow} O=N-O-N=O + H_2O$$

Dinitrogen dioxide

initially transforms the amino group into a nitrosamine. The nitrosamine equilibrates with the diazoic acid which then dehydrates to the diazonium salt (7).

(6) Sulfanilic acid   **(Nitrosamine)**   **(7) Diazonium cation**

This ion reacts at the para position of the electron-rich ring of N,N-dimethylaniline (8) by a coupling reaction that is mechanistically an electrophilic substitution reaction. The para position is preferred since it is sterically less crowded. (Attack at ortho positions are possible in substrates where the para position is sterically hindered.) Under the acid conditions of the reaction, a bright red acid, helianthin (9), is formed. Helianthin is converted by base to the orange sodium salt, methyl orange (5).

(7)   (8)

(9) Helianthin (red)

(5) Methyl orange (orange)

Typical of a dye, methyl orange contains a binding group in the molecule, the sulfonic acid group, $-SO_3H$. This is the group that binds the dye to the substrate being dyed. Other binding groups are the phenolic, Ar-OH, and carboxylic acid, $-CO_2H$, groups.

Methyl orange is often used for titrations where the end-point occurs in the range of pH 3.2 to 4.4. In dilute solution at a pH > 4.4, the negative ion predominates and the solution is yellow. At a pH of 3.2, it is protonated and the red helianthin form colors the solution. Other azo dyes, such as congo red and para red, can behave as an indicator in acid-base titrations. Like methyl orange, the structure changes by protonation and deprotonation at various pH values and different colors result.

OBJECTIVES

1. To generate an arene diazonium salt.
2. To use the arene diazonium salt in a diazo coupling reaction to synthesize an azo dye.
3. To demonstrate the ability of the azo dye to act as an indicator.

PROCEDURE

CAUTION!   *N,N-Dimethylaniline is toxic. Avoid breathing the vapor; dispense in the hood. The chemical is absorbed through the skin; avoid contact; if any comes into contact with your skin, wash thoroughly with soap and water; wear gloves when using. Methyl orange is an orange dye. Avoid touching the chemical; wear gloves to minimize dyeing your fingers. Wash your glassware thoroughly with acetone when finished.*

Formation of the diazonium salt of sulfanilic acid (4-diazobenzenesulfonic acid, sodium salt)

1. Weigh out 2.4 g of sulfanilic acid monohydrate. Record the weight to the nearest 0.01 g on the Report Sheet (1).

2. Place in a 125-mL Erlenmeyer flask the sulfanilic acid monohydrate and 25 mL of 2.5% aqueous sodium carbonate solution. Heat to a boil on a hot plate until all the solid dissolves.

3. Remove the solution from the hot plate, and while swirling the flask, cool in an ice-water bath to a temperature of 5 - 10°C.

4. Add 1.0 g of sodium nitrite to the solution and stir with a glass rod until it is dissolved.

5. Pour the solution into a 250-mL beaker containing 15 g of ice and 3 mL of concentrated hydrochloric acid and stir. A finely divided white suspension of the diazonium salt is formed and is ready for use. Since the diazonium salt is relatively stable, it will stay for a few hours. Keep the suspension in the ice-water bath until you need to use it.

Formation of methyl orange {4-[4-(dimethylamino)phenylazo]benzenesulfonic acid, sodium salt}

1. Pipet 1.6 mL of N,N-dimethylaniline and 1.3 mL of glacial acetic acid into a test tube (13 × 100 mm). Mix the two liquids well by using a Pasteur pipet: draw up the liquid into the pipet and squirt it out into the test tube; repeat until mixed. *Dispense in the hood.* (Remember to wear gloves from here on!)

2. Add the solution of dimethylaniline acetate, dropwise, with a Pasteur pipet, to the cooled suspension of diazotized sulfonic acid contained in the 250-mL beaker; stir with a glass rod as you add the solutions together. Rinse the test tube with 5 mL of water and add to the beaker. Keep stirring the solution for an additional 10 min. while it cools in the ice-water bath. A paste forms.

3. Add 20 mL of 10% sodium hydroxide solution, dropwise, with a Pasteur pipet. As you add the base, continue to stir with a glass rod and to cool in the ice-water bath. With the glass rod take a drop of solution and test with pH paper; the solution should be basic (you may need to take a drop and dilute with water in a test tube before testing with pH paper). If the solution is still acidic, continue to add base, testing after each 0.5 mL addition, until the solution is basic. The orange sodium salt should form.

4. Stir well as the mixture is heated on the hot plate to boiling. Most of the dye dissolves along with salt impurities. Place the beaker in an ice-water bath and allow the solution to cool.

5. When thoroughly cooled, collect the product by vacuum filtration using a small Büchner (or Hirsch funnel) (see Fig. 15.1, pg. 156, for technique). Rinse the beaker with two 10 mL portions of saturated aqueous sodium chloride solutions and wash the filter cake in the Büchner funnel with these rinse solutions (the filtrate will be dark).

Crystallization

1. The crude, wet product can be crystallized at once from water.

2. Transfer the filter cake, along with the filter paper, to a 250-mL beaker containing 100 mL of boiling water. Keep a gentle boil, and stir with a glass rod for a few minutes. Be careful not to destroy the filter paper since you want to remove it once the dye has dissolved. Once most of the dye has gone into solution, remove the paper and allow the solution to cool to room temperature. Then cool in an ice bath.

3. Filter the cold solution by vacuum using a small Büchner funnel. Allow the solid to air dry by keeping the water aspirator going and allowing air to flow through the filter cake.

4. Carefully separate the methyl orange crystals from the filter paper. Spread on a watch glass and allow the product to dry until the next period.

5. Weigh the dried methyl orange crystals to the nearest 0.01 g and record on the Report Sheet (3). Calculate the percent yield (5). A melting point for the salt need not be taken.

6. Place the crystals in a vial with a neat label containing your name, the name of the product and the percent yield.

7. Waste solutions should be collected in an appropriate container. All glassware should be thoroughly washed with soap and water and rinsed with acetone. Any orange residue will contaminate future experiments if equipment is not clean.

Indicator test

1. Place a small sample of the methyl orange crystals in a test tube (13 × 100 mm). Add 5 mL of water.

2. With a Pasteur pipet add 2 drops of dilute hydrochloric acid; record the color of the solution on the Report Sheet (6).

3. Carefully add dilute sodium hydroxide, dropwise, until basic; record the color of the solution on the Report Sheet (7).

CHEMICALS AND EQUIPMENT

1. N,N-Dimethylaniline
2. Glacial acetic acid
3. Sulfanilic acid monohydrate
4. Concentrated hydrochloric acid, HCl
5. 1 M hydrochloric acid, 1 M HCl
6. 2.5% aqueous sodium carbonate, Na_2CO_3
7. Saturated aqueous sodium chloride, NaCl
8. 10% aqueous sodium hydroxide, NaOH
9. 1 M sodium hydroxide, 1 M NaOH
10. Sodium nitrite, $NaNO_2$
11. Büchner funnel, 50 mm OD (or Hirsch funnel)
12. Filter flask, 250-mL
13. Filter paper, Whatman no. 1, 4.25 cm
14. Hot plate
15. Pasteur pipets
16. pH paper (range 1 - 12)
17. Pipet, 2-mL graduated
18. Spectroline pipet filler
19. Test tube (13 × 100 mm)
20. Vacuum tubing, 2 ft. length
21. Vial

EXPERIMENT 23

NAME _____ SECTION _____ DATE _____

PARTNER _____ GRADE _____

PRE-LAB QUESTIONS

1. Methyl orange is an example of a family of dyes referred to as *azo dyes*. Explain why this term applies to this compound.

2. Methyl orange, congo red and para red are used as acid-base indicators. How can these compounds be used in an acid-base titration?

3. Why are azo dyes colored?

4. What expression best describes the mechanism for this reaction?

EXPERIMENT 23

NAME _____   SECTION _____   DATE _____

PARTNER _____   GRADE _____

REPORT SHEET

1. Weight of sulfanilic acid monohydrate   _____ g

2. Moles of sulfanilic acid monohydrate   _____ moles
    [(1)/191.19]

3. Weight of methyl orange   _____ g

4. Moles of methyl orange   _____ moles
    [(3)/327.34]

5. Percent yield   _____ %

    $$\% \text{ Yield } = \frac{\text{Moles of methyl orange (4)}}{\text{Moles of sulfanilic acid monohydrate (2)}} \times 100 =$$

6. Color in acid

7. Color in base

POST-LAB QUESTIONS

1. Use equations to show how N,N-dimethylaniline (8) is activated at the para position. Why does the substitution take place at the para position and not at the ortho position?

2. Below is the structure of FD & C Yellow No. 6 (Sunset yellow), a food color approved by the Food and Drug Administration in 1975. Trace the extended conjugated system in the molecule.

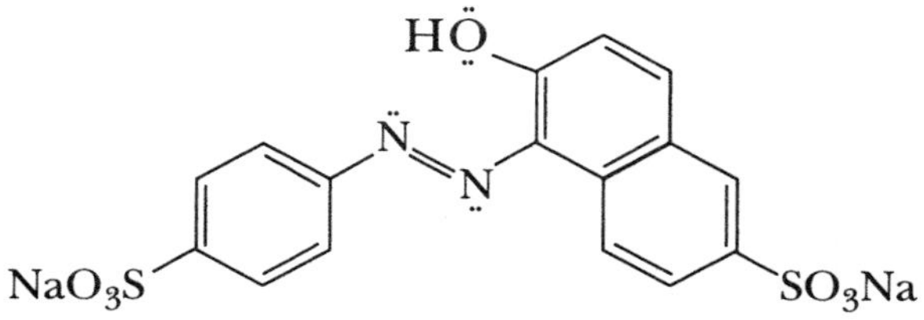

FD & C Yellow No. 6

3. Below is the structure of methyl red (10). Draw the structure of the amines necessary to synthesize this dye.

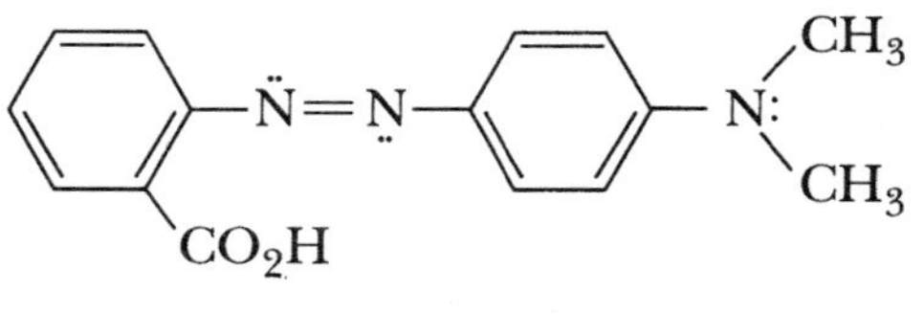

(10) Methyl red

4. What is the binding group in congo red (3)? In methyl red (10)?

Isocitrate Dehydrogenase: an Enzyme of the Citric Acid Cycle

BACKGROUND

The citric acid cycle is the first unit of the common metabolic pathway through which most of our food is oxidized to yield energy. In the citric acid cycle the partially fragmented food products are broken down further. The carbons of the C_2 fragments are oxidized to CO_2, released as such and expelled in the respiration. The hydrogens and the electrons of the C_2 fragments are transferred to the coenzyme, nicotinamide adenine dinucleotide, NAD^+, or to flavin adenine dinucleotide, FAD, which in turn become $NADH + H^+$ or $FADH_2$, respectively. These enter the second part of the common pathway, oxidative phosphorylation, and yield water and energy in the form of ATP.

The first enzyme of the citric acid cycle to catalyze both the release of one carbon dioxide and the reduction of NAD^+ is isocitrate dehydrogenase. The overall reaction of this step:

$$
\begin{array}{l}
\text{COO}^- \\
|\\
\text{CH}_2 \\
|\\
\text{CH}\!-\!\text{COO}^- \\
|\\
\text{HO}\!-\!\text{CH} \\
|\\
\text{COO}^-
\end{array}
\; + \; NAD^+ \xrightarrow[\text{enzyme}]{}
\begin{array}{l}
\text{COO}^- \\
|\\
\text{CH}_2 \\
|\\
\text{CH}_2 \\
|\\
\text{C}\!=\!\text{O} \\
|\\
\text{COO}^-
\end{array}
\; + \; NADH \; + \; CO_2
$$

Isocitrate α-**Ketoglutarate**

The reduction of the NAD^+ itself is given by the equation:

$$
\text{NAD}^+ \;+\; H^+ \;+\; 2\,e^- \; \rightleftharpoons \; \text{NADH}
$$

NAD$^+$ **NADH**

The enzyme has been isolated from many tissues, the best source being a heart muscle or yeast. The isocitrate dehydrogenase requires the presence of cofactors Mg^{2+} or Mn^{2+}. As an allosteric enzyme it is regulated by a number of modulators. ADP, adenosine diphosphate, is a positive modulator and therefore, stimulates enzyme activity. The enzyme has an optimum pH of 7.0.

235

As is the case with all enzymes of the citric acid cycle, isocitrate dehydrogenase is found in the mitochondria. To isolate the enzyme, the cells of the yeast must be disintegrated which can be done by grinding them with sand.

In the present experiment you will determine the activity of isocitrate dehydrogenase extracted from baker's yeast. The basis of the measurement of the enzyme activity is the absorption spectrum of NADH. This reduced coenzyme has an absorption maximum at 340 nm. Therefore, an increase in the absorbance at 340 nm indicates an increase in NADH concentration, hence the progress of the reaction. We define the unit of isocitrate dehydrogenase activity as one that causes an increase of 0.01 absorbance per minute at 340 nm.

For example, if a 10 mL solution containing isocitrate and yeast extract exhibits a 0.04 change in the absorbance in two minutes, the enzyme activity will be

$$\frac{0.04}{2 \times 10 \text{ mL}} \times \frac{1 \text{ unit}}{0.01/1} = 0.2 \text{ units/mL}$$

OBJECTIVES

To isolate an enzyme of the citric acid cycle, isocitrate dehydrogenase, and to measure its activity.

PROCEDURE

A. **Enzyme extraction.** It is preferable to work in pairs. If not enough mortars and pestles are available four students can work together in preparing the enzyme extract. In the latter case, double the amounts recommended. Cool a mortar in ice water. Weigh 2 g of fresh baker's yeast and place it in the mortar. Record the weight of the yeast on the Report Sheet (1). Add approximatly 4 g of acid washed sand. Grind the yeast and sand with a pestle for 10 min. to break up the cells. Add 5 mL ice cold 0.1 M NaHCO₃ solution. Record the volume on your Report Sheet (2). Continue grinding with the pestle vigorously for five min. Allow the sand and the broken yeast fragments to settle out of the solution for five min.

B. **Light absorption measurements.** This part can also be performed in pairs to reduce the number of spectrophotometers needed. The detailed instructions how to operate the spectrophotometer depends on the make of the instrument and will be provided by your lab instructor. The general outlines of the operation are given below:

1. While waiting for the sand to settle in the mortar, turn on the instrument and let it warm up for a few minutes. Turn the wavelength control knob to read 340 nm. With no sample tube in the sample compartment, adjust the amplifier control knob so that 0% transmittance or infinite absorbance is read.

2. Fill a sample tube with distilled water and insert it into the spectrophotometer. Adjust the reading to 100% transmittance (or 0 absorbance). This zeroing must be performed every 5 - 10 min. since some instruments have the tendency to drift. The instrument is now ready to measure enzyme activity.

3. Decant the yeast extract from the mortar into a clean test tube labelled "extract". If your extract looks brown and turbid you must purify it by vacuum filtration (Fig. 24.1).

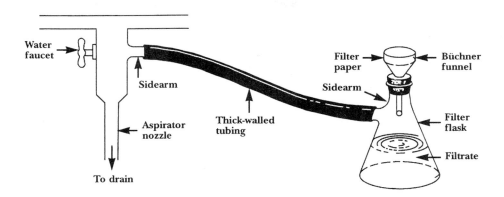

Figure 24.1 Vacuum filtration apparatus.

Place a Whatman No.1 filter paper in the Büchner funnel. Wet it with a few mL of distilled water and apply suction by opening the water faucet. The filter paper must fit snugly and adhere to the Büchner funnel. Weigh about 1 g of Celite filter aid in a 50-mL beaker. Add 25 mL distilled water and make a slurry by stirring it. While the Büchner funnel is under suction pour the slurry on top of the filter paper. Let it filter until the Celite appears to be dry. While the water is still running *disconnect* the suction flask from the faucet by *removing the rubber hose from the side arm of the suction flask*. With this maneuver you prevent the tap water entering into the suction flask. Remove the Büchner funnel and empty the water from the suction flask. Rinse it with distilled water and empty it again. Insert the Büchner funnel back into the suction flask and reconnect it to the aspirator. While the Büchner funnel is under suction, pour your yeast extract on top of the Celite filter aid. The filtrate coming through should be clear or only slightly turbid, otherwise the entire filtration procedure must be repeated. When the Celite filter aid appears to be dry your filtration is completed. Disconnect the rubber hose from the side arm of the suction flask. Turn off the faucet. Transfer the filtered yeast extract into a clean and dry 10-mL graduate cylinder and measure its volume. Report it on the Report Sheet (3).

4. Prepare two test tubes for the enzyme activity measurements. Add 0.4 mL of phosphate buffer to *each* test tube. Next add to each test tube 0.2 mL $MgCl_2$, 0.5 mL ADP and 0.5 mL NAD^+ solutions. Mix the contents of the test tubes. Label one of the test tubes "Blank" and add to it 1.0 mL yeast extract and 2.2 mL distilled water. Read its transmission (absorbance) in the spectrophotometer at 340 nm, and record this value on the Report Sheet (6).

5. Label the second test tube "Sample". Add to it 1.0 mL yeast extract. Record this value on the Report Sheet (4). Add 1.2 mL isocitrate solution. *Note the time of the isocitrate addition. This is zero time.* Add also 1.0 mL distilled water and mix the contents. Record the total volume on the Report Sheet (5). Read the transmission (absorbance) of the "Sample" test tube 30 sec. after the addition of isocitrate and again after 60, 90 120, 150 and 180 sec. Take a last reading 5 min. after zero time. Record your observations.

6. If your enzyme activity was low and you did not get 0.1 - 0.5 absorbance increase during the 3 min. period, repeat the experiment as in step 5 using greater amount of enzyme extract. For example, if you use 2 mL enzyme extract in each test tube, you then should add only 1.2 mL distilled water to your "Blank" test tube and none to your "Sample" test tube.

CHEMICALS AND EQUIPMENT

1. Baker's yeast
2. Sand
3. $NaHCO_3$ buffer
4. Phosphate buffer, pH 7.0
5. 2.5 mM ADP solution
6. 2.0 mM NAD^+ solution
7. 5.0 mM isocitrate solution
8. Celite filter aid
9. Büchner funnel
10. Filter flask
11. Filter paper

EXPERIMENT 24

NAME _____ SECTION _____ DATE _____

PARTNER _____ GRADE _____

PRE-LAB QUESTIONS

1. What reaction precedes the isocitrate oxidation (dehydrogenation) in the citric acid cycle?

2. What reaction follows the isocitrate oxidation (dehydrogenation) in the citric acid cycle?

3. How do we measure isocitrate dehydrogenase activity?

4. What is the function of sand in the enzyme preparation?

EXPERIMENT 24

NAME _____ SECTION _____ DATE _____

PARTNER _____ GRADE _____

REPORT SHEET

1. Weight of baker's yeast   _____ g
2. Volume of $NaHCO_3$ added   _____ mL
3. Volume of yeast extract after filtration   _____ mL
4. Volume of the yeast extract in "Sample"   _____ mL
5. Total volume of the "Sample"   _____ mL

Net absorbance of sample = absorbance of sample - absorbance of blank.

6. Absorbance of blank   _____

Time (s)	Absorbance of sample	Net absorbance of sample
30		
60		
90		
120		
150		
180		
400		

1. Plot your data: Net absorption versus time. At zero time the net absorption is zero.

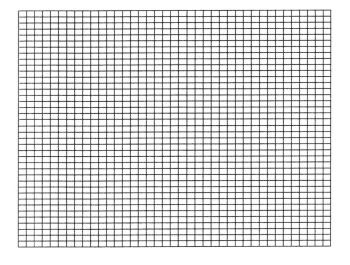

2. Calculate the enzyme activity:

    (a) Units of enzyme activity/ reaction mixture: The initial slope of the plot is usually a straight line. If so, read the value of net absorbance at 60 sec. Divide it by 0.01. This gives you the number of enzyme activity units/reaction mixture. (One unit of enzyme activity is 0.01 net absorbance/min.)

    (b) Calculate the isocitrate dehydrogenase activity per mg yeast.

    Example: If the net absorbance/min. was 0.04. This is 4 units of activity. You used 0.2 mL of yeast extract from a total volume of 5 mL. This volume came from the original 2 g of baker's yeast.

$$\frac{4 \text{ units}}{0.2} \times \frac{5}{2} \times \frac{1}{1000 \text{ mg}} = 5 \times 10^{-2} \text{ units/ mg yeast}$$

POST-LAB QUESTIONS

1. What kind of enzyme activity would you observe if you did not grind the yeast sufficiently and only 10% of the cells were broken?

2. Isocitrate dehydrogenase has an optimum pH of 7.0. Did you work under optimal condition? If so, how did you provide for it?

3. Inspect your plot. Why didn't you take the absorbance at 5 min. rather than at 60 sec. to calculate the enzyme activity? If you would have used the 5 min. value would your enzyme activity be higher, lower or the same as the one you calculated? Explain.

Preparation of a Hand Cream

BACKGROUND

Hand creams are formulated to carry out a variety of cosmetic functions. Among these are: softening the skin and preventing dryness; elimination of natural waste products (oils) by emulsification; cooling the skin by radiation, thus helping to maintain body temperature. In addition, hand creams must have certain ingredients that help spreadability and provide body to the hand cream. In many cases added fragrance improves the odor, and in some special cases medications combat assorted ills.

The basic hand cream formulations all contain water to provide moisture and lanolin to help its absorption by the skin. The latter is a yellowish wax. Chemically wax is made of esters of long chain fatty acids and long chain alcohols. Lanolin is usually obtained from sheep wool; it has the ability to absorb 25 - 30% of its own weight of water and to form a fine emulsion. Mineral oil, which consists of high molecular weight hydrocarbons, provides spreadability. In order to allow nonpolar substances, such as lanolin and mineral oil, to be uniformly dispersed in a polar medium, water, one needs strong emulsifying agents. An emulsifying agent must have nonpolar, hydrophobic portions to interact with the oil and also polar, hydrophilic portions to interact with water. A mixture of stearic acid and triethanolamine, through acid-base reaction, yields the salt that has the requirements to act as an emulsifying agent.

Beside these five basic ingredients some hand creams contain alcohols such as propylene glycol (1,2-propanediol) and esters such as methyl stearate to provide the desired texture of the hand cream.

In this experiment you will prepare four hand creams using combination of ingredients as shown in Table 25.1.

OBJECTIVES

1. To learn the method of preparing a hand cream
2. To appraise the function of the ingredients in the hand cream.

PROCEDURE

Preparation of the Hand Creams

For each sample in Table 25.1 assemble the ingredients in two beakers. Beaker 1 contains the polar ingredients and Beaker 2 the nonpolar contents.

TABLE 25.1

Ingredients	Sample 1	Sample 2	Sample 3	Sample 4	
Water	25 mL	25 mL	25 mL	25 mL	
Triethanolamine	1 mL	1 mL	1 mL	–	Beaker1
Propylene glycol	0.5 mL	0.5 mL	–	0.5 mL	
Stearic acid	5 g	5 g	5 g	5 g	
Methyl stearate	0.5 g	0.5 g	–	0.5 g	
Lanolin	4 g	4 g	4 g	4 g	Beaker 2
Mineral oil	5 mL	–	5 mL	5 mL	

1. To prepare sample 1 put the nonpolar ingredients in a 50-mL beaker (beaker 2) and heat it in a water bath. The water bath can be a 400-mL beaker half filled with tap water and heated with a Bunsen burner (Fig. 25.1). Carefully hold the beaker with crucible tongs in the boiling water until all ingredients melt.

2. In the same water bath heat the 100-mL beaker (beaker 1) containing the polar ingredients for about 5 min. Remove the beaker and set it on the bench top.

3. Into the 100-mL beaker containing polar ingredients pour slowly the contents of the 50-mL beaker that holds the molten nonpolar ingredients (Fig. 25.2). Stir the mixture for 5 min. until you have a smooth uniform paste.

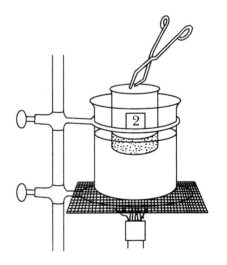

Figure 25.1 Heating ingredients.

Figure 25.2 Mixing hand cream ingredients.

4. Repeat the same procedure in preparing the other three samples.

Characterization of the Hand Cream Preparations

1. Test the pH of the hand creams prepared using a wide range pH paper.

2. Rubbing a small amount of the hand cream between your fingers test for smoothness and homogeneity. Also note the appearance.

CHEMICALS AND EQUIPMENT

1. Bunsen burner
2. Lanolin
3. Stearic acid
4. Methyl stearate
5. Mineral oil
6. Triethanolamine
7. Propylene glycol
8. Crucible tongs

EXPERIMENT 25

NAME _____ SECTION _____ DATE _____

PARTNER _____ GRADE _____

PRE-LAB QUESTIONS

1. What is the function of lanolin in the hand cream?

2. The emulsifying agent was prepared from stearic acid and triethanolamine. Give the name of this salt. Write its formula.

3. What functional groups of the emulsifying agent provide the hydrophilic character?

EXPERIMENT 25

NAME _____ SECTION _____ DATE _____

PARTNER _____ GRADE _____

<u>REPORT SHEET</u>

Characterization of the Hand Cream Samples

<u>Properties</u>	<u>Sample 1</u>	<u>Sample 2</u>	<u>Sample 3</u>	<u>Sample 4</u>
pH				
Smoothness				
Homogeneity				
Appearance				
Missing ingredient				

POST-LAB QUESTIONS

1. In comparing the properties of the hand creams you produced, ascertain what is the function of each of the missing ingredients in the hand cream:

    (a) Mineral oil

    (b) Triethenolamine

    (c) Methyl stearate and propylene glycol

2. A hand cream appears smooth and uniform after you prepared it, but in a week of storage most of the water settles on the bottom and most of the oil separates on the top. What do you think may have gone wrong with the hand cream preparation?

3. Could you prepare a hand cream without water? Would it serve the cosmetic functions listed in the "Background" section?

Isolation of Lipids from Egg Yolk

BACKGROUND

Lipids are a group of chemicals that are characterized by their insolubility in water. The lipids are divided into three classes: simple lipids (fats, oils, and waxes), complex lipids (phospholipids, glycolipids, and sphingolipids), and steroids. These compounds can be separated from each other on the basis of their solubility in different organic solvents.

Egg yolk has a rich supply of lipids. The two most prominent lipids in the egg yolk are cholesterol, which is a steroid, and phosphatidyl choline (or lecithin by its common name), which is a phospholipid.

Cholesterol

Lecithin

In this experiment, we isolate these two compounds. The egg yolk also contains smaller amounts of neutral fat, glycolipids, and sphingolipids. These may appear as contaminants in your preparations. We do not aim to isolate these minor constituents. The basis for the isolation of cholesterol and lecithin is their solubility in acetone and diethyl ether. Cholesterol is soluble in acetone, lecithin is not. First, we extract the cholesterol with acetone. Next, we extract the lecithin from the residue of the egg yolk by using diethyl ether as a solvent. Other nutrients (carbohydrates and proteins) are left behind in the residue.

The extracts are not pure compounds. They contain minor lipid constituents. We purify the cholesterol by crystallization from its solution. The contaminant lipid components will stay in solution. We purify the lecithin by precipitating it from the diethyl ether solution. This is accomplished by adding acetone to the solution. The contaminating minor lipid constituents will stay in the diethyl ether-acetone solution. Lecithin and cholesterol are both important constituents of the membranes of cells. They are found in high concentration in the brain and nerves. Cholesterol is also found in the blood. It serves as raw material for the synthesis of many steroids. When the cholesterol concentration increases and the amount of the lipoprotein, the carrier of cholesterol in the blood, is insufficient, cholesterol may be deposited in the blood vessels in the form of plaques (artherosclerosis). Excess cholesterol also forms gallstones.

Lecithin, having a strong polar end and long nonpolar tails of fatty acids, is a good emulsifying agent, dispersing fat in water. The food industry uses lecithin as an emulsifying agent in many products.

OBJECTIVES

1. To illustrate the differential solubility of cholesterol and lecithin.
2. To isolate and purify these compounds on the basis of this property alone.

PROCEDURE

Be absolutely certain that no flames (Bunsen burner, matches) are used during this lab period. The solvents employed are flammable and the vapors of the solvents can be ignited by an open flame. Only hot plates are allowed to heat steam baths or, if live steam is available, use that directly for the steam bath. Make certain that when you disconnect the hot plate, you pull the plug and not the cord from the outlet. Pulling the cord occasionally may create sparks, which may ignite the vapors. Place your hot plate and steam bath in the hood and bring the water in the steam bath to a boil.

Extraction

1. Take one half the yolk of a hard boiled egg. Record its weight. Mash it to a smooth consistency with a spatula in a 250-mL beaker. Add 25 mL acetone. Continue to stir the mixture for 5 min. Allow the solid to settle and decant the acetone extract containing the crude cholesterol. Pour the extract into a labeled 100-mL Erlenmeyer flask and stopper it with a cork. Repeat the extraction with another portion of 25 mL of acetone. Combine the two acetone extracts.

2. Do the next operation under the hood. Evaporate the remaining acetone from the egg yolk residue by putting the beaker on the steam bath. *If the evaporation proceeds too fast some of the egg yolk residue may splatter. You can avoid this by removing your beaker occasionaly from the steam bath.* The whole evaporation will take a few minutes. Remove the beaker and let it cool. Add 15 mL of diethyl ether (under the hood) to the egg yolk and extract the lecithin by stirring vigorously for 2 min. Decant the ether extract and pour it into a labeled 100-mL Erlenmeyer flask and cork it. Repeat the extraction with another portion of 15-mL diethyl ether. Decant and combine the two extracts. Place the residual egg yolk in a wide mouth jar labeled "Waste."

Isolation and Purification

1. Filter the acetone extract into a clean 100-mL beaker. Evaporate the acetone by placing the beaker on the steam bath in the hood until about 10 mL of extract is left. Transfer this warm acetone extract into a clean and dry 50-mL Erlenmeyer flask. Close the top with a cork. Place the Erlenmeyer flask into an ice bath and wait for 20 min. A white precipitate will form. This is your crude cholesterol preparation.

2. While waiting for the cholesterol to precipitate, you can isolate the lecithin from the ether solution. Filter the combined ether extracts into a preweighed 100-mL beaker. Place the beaker on the steam bath **under the hood** and evaporate most of the diethyl ether until you are left with about 10 mL of ether extract. Cool the beaker to room temperature. Add 30 mL of acetone to the beaker with continuous stirring. A waxy precipitate will form. Decant the solvent and discard it into the jar labeled "Waste." Evaporate the residual solvent by placing the beaker on the steam bath until the precipitate appears to be dry. *As before, during the evaporation, splattering may occur which you can avoid by removing the beaker from the steam bath occasionally.* Cool the beaker to room temperature and weigh it. Calculate the yield of lecithin.

3. Take the Erlenmeyer flask from the ice bath. Decant most of the cold solvent. Dissolve the cholesterol precipitate in 15 mL of acetone at room temperature. Stir it. Not all the precipitate will dissolve. The contaminating phospholipids will not dissolve. Filter the dissolved cholesterol solution into a preweighed watch glass. Allow the acetone to evaporate. The white crystals that formed are your cholesterol preparation. Weigh the watch glass. Calculate the yield of the isolated cholesterol.

    Preserve both the cholesterol and the lecithin preparations for your next experiment (Exp. 27).

CHEMICALS AND EQUIPMENT

1. Hard boiled egg
2. Acetone
3. Diethyl ether
4. Hot plate
5. Steam bath
6. Waste jar

EXPERIMENT 26

NAME _____ SECTION _____ DATE _____

PARTNER _____ GRADE _____

PRE-LAB QUESTIONS

1. Identify the polar group(s) in cholesterol.

2. Acetone can be mixed with water in any proportion. Diethyl ether is insoluble in water. Which one is a more polar solvent, acetone or diethyl ether? Explain.

3. Which part of the lecithin molecule contributes its nonpolar characteristic?

4. Why are you not allowed to use an open flame (Bunsen burner) during this experiment?

EXPERIMENT 26

NAME _____ SECTION _____ DATE _____

PARTNER _____ GRADE _____

<u>REPORT SHEET</u>

1. Weight of the paper (tare)   _____ g

2. Weight of the egg yolk + paper   _____ g

3. Weight of the egg yolk: (2) − (1)   _____ g

4. Weight of beaker   _____ g

5. Weight of lecithin + beaker   _____ g

6. Weight of lecithin: (5) − (4)   _____ g

7. % yield of lecithin = (weight of lecithin/
   weight of yolk) × 100 = [(6)/(3)] × 100   _____ %

8. Appearance of lecithin: _____

9. Weight of watch glass   _____ g

10. Weight of watch glass + cholesterol   _____ g

11. Weight of cholesterol: (10) − (9)   _____ g

12. % yield of cholesterol = (weight of cholesterol/weight of yolk)
    × 100 = [(11)/(3)] × 100   _____ %

13. Appearance of cholesterol: _____

POST-LAB QUESTIONS

1. What is the difference in appearance between cholesterol and lecithin?

2. Consumer charts indicate that one egg yolk contains 250 mg of cholesterol. Compare this to your yield. Assuming that your preparation is pure cholesterol, what percent of the available cholesterol did you isolate?

3. The normal cholesterol level in the blood is 1.0 mg/mL. Your body contains 5.5 L of blood. You just ate two hard boiled eggs. Assuming that all the cholesterol from the yolk is absorbed into your bloodstream, calculate what will be the cholesterol concentration in your blood after the meal.

4. Why does cholesterol, in the absence of protective lipoproteins, form plaques in the blood vessels?

Analysis of Lipids

EXPERIMENT 27

BACKGROUND

Lipids are chemically heterogeneous mixtures. The only common property they have is their insolubility in water. We can test for the presence of various lipids by analyzing their chemical constituents. In Experiment 26 you isolated the lipids cholesterol and lecithin from egg yolk. These may be pure compounds or they still may contain other lipids as contaminants.

Structurally, cholesterol is completely different from lecithin. It contains a five-membered fused ring that is the common core of all steroids.

Steroid nucleus   **Cholesterol**

There is a special colorimetric test, the Lieberman-Burchard reaction, which uses acetic anhydride and sulfuric acid as reagents, that gives a characteristic green color in the presence of cholesterol. This color is due to the -OH group of cholesterol and the unsaturation found in the adjacent fused ring. The color change is gradual: first it appears as a pink coloration, changing later to lilac, and finally to deep green.

Lecithin is a complex lipid; it contains a number of components that can be detected and thereby shows the presence of this phospholipid. When lecithin is hydrolyzed in acidic medium, both the fatty-acid ester bonds and the phosphate ester bonds are broken and free fatty acids and inorganic phosphate are released.

Using a molybdate test, we can detect the presence of phosphate in the hydrolysate by the appearance of a purple color. Although this test is not specific for lecithin (other phosphate containing

255

lipids will give a positive molybdate test), it differentiates clearly between cholesterol (negative test) and phospholipids (positive test).

A second test that differentiates between cholesterol and lecithin is the acrolein reaction. When lipids containing glycerol are heated in the presence of potassium hydrogen sulfate the glycerol is dehydrated, forming acrolein, which has an unpleasant odor. Further heating results in polymerization of acrolein, which is indicated by the slight blackening of the reaction mixture. Both the pungent smell and the black color indicate the presence of glycerol and thereby lecithin. Cholesterol gives a negative acrolein test.

$$
\begin{array}{ccc}
CH_2OH & & CH_2 \\
| & & \parallel \\
CHOH & \xrightarrow{\Delta} & CH \quad + \quad 2H_2O \\
| & & | \\
CH_2OH & & C=O \\
& & | \\
& & H
\end{array}
$$

The purity of both the lecithin and cholesterol preparations can be tested by melting point measurements. A pure compound gives a sharp melting point within $\pm 1\,°C$ range. Compounds that are contaminated by the presence of other constituents give a relatively broad melting point range of $\pm 5\,°C$. Impurities usually lower the melting point of a compound. Impure lecithin may even decompose before reaching the melting point. By taking the melting points of your cholesterol and lecithin preparations and comparing them with the literature values of the pure compounds, you can get a good idea regarding the purity of your compounds.

OBJECTIVES

To investigate the purity of cholesterol and lecithin preparations by using colorimetric tests and melting point measurements.

PROCEDURE

In Experiment 26 you isolated two compounds from egg yolk: cholesterol and lecithin. We will test these compounds for their purity.

Phosphate test

1. Add about 0.2 g of cholesterol to one clean and dry test tube and the same amount of lecithin to another. Hydrolyze the compounds by adding 3 mL of 6 M nitric acid to each test tube.

CAUTION!   *6 M nitric acid is a strong acid. Handle it with care. Use gloves.*

2. Prepare a water bath by boiling about 100 mL of tap water in a 250-mL beaker on a hot plate. Place the test tubes in the boiling water bath for 5 min. Do not inhale the vapors. Cool the test tubes. Neutralize the acid by adding 3 mL of 6 M NaOH. Mix.

3. Transfer 2 mL of each neutralized sample into clean and labeled test tubes. Add 3 mL of a molybdate solution to each test tube and mix the contents. *(Be careful. The molybdate solution contains sulfuric acid.)* Heat the test tubes in a boiling water bath for 5 min. Cool them to room temperature.

4. Add 0.5 mL of an ascorbic acid solution and mix the contents thoroughly. Wait for 20 min. for the development of the purple color. While you wait you can perform the rest of the colorimetric tests.

The acrolein test for glycerol

1. Place 1 g potassium hydrogen sulfate $KHSO_4$ in a clean, dry test tube and add a few crystals of your cholesterol preparation.

2. To a second test tube, add also 1 g of $KHSO_4$ and a few grains of your lecithin preparation. Set up your Bunsen burner in the hood. It is important that this test should be performed under the hood because of the pungent odor of the acrolein.

3. Gently heat each test tube, one at a time, over the Bunsen burner's flame, shaking it continuously from side to side. When the mixture melts, it slightly blackens and you will notice the evolution of fumes. Stop the heating. Smell the test tubes by moving them sideways under your nose or waft the vapors. Do not inhale the fumes directly. A pungent odor, resembling burnt hamburgers, is the positive test for glycerol. Do not overheat the test tubes, for the residue will become hard, making it difficult to clean the test tubes.

Lieberman-Burchard test for cholesterol

Place a few crystals of your cholesterol preparation in one labeled test tube. Add about the same amount of lecithin to a second clean and labeled test tube. *(The next step should be done in the hood)*. Transfer 3 mL of chloroform and 1 mL of acetic anhydride to each test tube. Finally, add one drop of concentrated sulfuric acid to each mixture. Mix the contents and record the color changes, if any. Wait 5 min. Record again the color of your solutions.

Melting point measurements

1. Place a few grains of cholesterol crystals into a capillary tube. Bring the crystals to the closed bottom end of the capillary by either using a file in a bowing motion across the capillary or dropping the capillary repeatedly through a 30 to 40 cm glass tube (see Fig. 4.2, pg. 39).

2. Place the capillary tube in a melting point measuring apparatus (see Figs. 4.3 and 4.4). Record the melting point of your cholesterol preparation. The <u>CRC Handbook of Chemistry and Physics</u> lists a value for the melting point of cholesterol as 148°C. Compare your result.

3. Repeat the melting point determination with your lecithin preparation. If your lecithin is waxlike, you may have to disperse your sample to a powdery consistency in a mortar with pestle before you can place it in your capillary. The <u>Handbook</u> gives the melting point of lecithin as 236°C.

CHEMICALS AND EQUIPMENT

1. 6 M NaOH
2. 6 M HNO_3
3. Molybdate reagent
4. Ascorbic acid solution
5. $KHSO_4$
6. Chloroform
7. Acetic anhydride
8. Sulfuric acid, H_2SO_4
9. Melting point apparatus
10. Capillary tubes
11. Hot plate

EXPERIMENT 27

NAME _____ SECTION _____ DATE _____

PARTNER _____ GRADE _____

PRE-LAB QUESTIONS

1. What is the difference in intermolecular interaction in lecithins vs. cholesterol that gives such a wide contrast in their melting points?

2. Cholesterol has an alcohol group. One could also dehydrate cholesterol (removing one water molecule by heating). Show what kind of structure you would expect from the dehydration of cholesterol.

3. List all the functional groups in (a) acrolein (b) choline.

4. What techniques can you use to get the cholesterol crystals to the bottom of the capillary tube?

EXPERIMENT 27

NAME _____ **SECTION** _____ **DATE** _____

PARTNER _____ **GRADE** _____

<u>REPORT SHEET</u>

<u>Tests</u>   <u>Cholesterol</u>   <u>Lecithin</u>

1. Phosphate
    a. Color
    b. Conclusions

2. Acrolein
    a. Odor
    b. Color
    c. Conclusions

3. Lieberman-Burchard
    a. Initial color
    b. Color after 5 min.
    c. Conclusion

4. Melting point
    Comments

POST-LAB QUESTIONS

1. What is your overall conclusion regarding the purity of your compounds?

2. Assume that your cholesterol was free from all other lipid contaminants. However, it was not completely dry; it still contained some acetone. How would that affect your tests?

   a. Acrolein:

   b. Lieberman-Burchard:

   c. Phosphate:

   d. Melting point:

3. Assume that the lecithin still contained all its ester linkages. Would you get a positive acrolein test? Explain.

4. A positive acrolein test is indicated by its odor as well as by its color. Which comes first? Explain.

Extraction and Identification of Fatty Acids from Corn Oil

BACKGROUND

Fats are esters of glycerol and fatty acids. Liquid fats are often called oils. Whether a fat is solid or liquid depends on the nature of the fatty acids. Solid animal fats contain mostly saturated fatty acids while vegetable oils contain high amounts of unsaturated fatty acids. To avoid arteriosclerosis, the hardening of arteries, diets are recommended which are low in saturated fatty acids as well as in cholesterol.

Note that even solid fats contain some unsaturated fatty acids, and oils contain saturated fatty acids as well. Beside the degree of unsaturation, the length of the fatty acid chain also influences whether a fat is solid or liquid. Short chain fatty acids, such as found in coconut oil, convey liquid consistency, inspite of the low unsaturated fatty acid content. Two of the unsaturated fatty acids, linoleic and linolenic acids, are essential fatty acids because the body cannot synthesize them from precursors; they must be included in the diet.

The four unsaturated fatty acids most frquently found in vegetable oils are:

Oleic acid:	$CH_3(CH_2)_7CH=CH(CH_2)_7COOH$
Linoleic acid:	$CH_3(CH_2)_4CH=CHCH_2CH=CH(CH_2)_7COOH$
Linolenic acid:	$CH_3CH_2CH=CHCH_2CH=CHCH_2CH=CH(CH_2)_7COOH$
Arachidonic acid:	$CH_3(CH_2)_4CH=CHCH_2CH=CHCH_2CH=CHCH_2CH=CH(CH_2)_3COOH$

All the C=C double bonds in the unsaturated fatty acids are cis double bonds which interrupt the regular packing of the aliphatic chains, and thereby, convey a liquid consistency at room temperature. This physical property of the unsaturated fatty acid is carried over to the physical properties of triglycerides (oils).

In order to extract and isolate fatty acids from corn oil, first, the ester linkages must be broken. This is achieved in the saponification reaction in which a triglyceride is converted to glycerol and the potassium salt of its fatty acids:

$$CH_2-O-\overset{\overset{O}{\|}}{C}-C_{17}H_{35}$$
$$CH-O-\overset{\overset{O}{\|}}{C}-C_{17}H_{35} \quad + \quad 3KOH \quad \longrightarrow \quad CHOH \quad + \quad 3C_{17}H_{35}\overset{\overset{O}{\|}}{C}-O^-K^+$$
$$CH_2-O-\overset{\overset{O}{\|}}{C}-C_{17}H_{35}$$

CH₂OH — CHOH — CH₂OH (Glycerol)

Triglyceride   **Glycerol**   **Potassium stearate**

In order to separate the potassium salts of fatty acids from glycerol, the products of the saponification mixture must be acidified. Subsequently, the fatty acids can be extracted by petroleum ether. To identify the fatty acids so isolated, they must be converted to their methyl ester by a perchloric acid catalyzed reaction:

$$C_{17}H_{35}COOH + CH_3OH \xrightarrow{\ HClO_4\ } C_{17}H_{35}\overset{\overset{\displaystyle O}{\displaystyle \|}}{C}-O-CH_3 + H_2O$$

The methyl esters of fatty acids can be separated by thin layer chromatography (TLC). They can be identified by comparing their rate of migration (R_f values) to the R_f values of authentic samples of methyl esters of different fatty acids (Fig. 28.1).

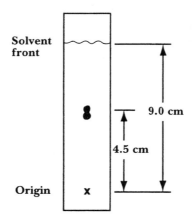

Figure 28.1 TLC chromatogram

R_f = distance travelled by fatty acid/distance travelled by the solvent front.

OBJECTIVES

1. To extract fatty acids from neutral fats.
2. To convert them to their methyl esters.
3. To identify them by thin layer chromatography.

PROCEDURES

Part A. Extraction of fatty acids

1. Weigh a 50-mL Erlenmeyer flask and record the weight on your Report Sheet (1).

2. Add 2 mL corn oil and weigh it again. Record the weight on your Report Sheet (2).

3. Add 5 mL of 0.5 M KOH in ethanol to the Erlenmeyer flask. Stopper it. Place the flask in a water bath at 55°C for 20 min.

4. When the saponification is completed, add 2.5 mL concentrated HCl. *(Caution: Strong Acid)*. Mix it by swirling the Erlenmeyer flask. Transfer the content into a 50-mL separatory funnel. Add 5 mL petroleum ether. Mix it thoroughly (see Fig. 6.1). Drain the lower aqueous layer into a flask and the upper petroleum ether layer into a glass stoppered test tube. Repeat the process by adding back the aqueous layer into the separatory funnel and extracting it with another portion of 5 mL petroleum ether. Combine the petroleum ether extracts.

Part B. Preparation of methyl esters

1. Place a plug of glass wool into the upper stem of a funnel, fitting it loosely. Add 10 g of anhydrous Na_2SO_4. Rinse the salt on to the glass wool with 5 mL petroleum ether; discard the wash. Pour the combined petroleum ether extracts into the funnel and collect the filtrate in an evaporating dish. Add another portion (2 mL) of petroleum ether to the funnel and collect this wash also in the evaporating dish.

2. Evaporate the petroleum ether under the hood by placing the evaporating dish on a water bath at 60°C. (Alternatively, if dry N_2 is available, the evaporation could be achieved by bubbling nitrogen through the extract. This also must be done under the hood).

3. When dry, add 10 mL CH_3OH:$HClO_4$ mixture (95:5). Place the evaporating dish in the water bath at 55°C for 10 min.

Part C. Identification of fatty acids

1. Transfer the methyl esters prepared above into a separatory funnel. Extract twice with 5 mL petroleum ether. Combine the petroleum ether extracts.

2. Prepare another funnel with anhydrous Na_2SO_4 on top of glass wool. Filter the combined petroleum ether extracts through the salt into a dry clean evaporating dish. Evaporate the petroleum ether on the water bath at 60°C as before. When dry, add 0.2 mL petroleum ether and transfer the solution to a dry and clean test tube.

3. Take a 15 × 6.5 cm TLC plate. Make sure you do not touch the TLC plate with your fingers. Preferably use plastic gloves, or handle the plate by holding it only at the edges. This precaution must be observed throughout the whole operation because your fingers may contaminate the sample. With a pencil lightly draw a line parallel to the 6.5 edge about 1 cm from the edge. Mark the positions of 5 spots, equally spaced, where you will spot your samples. (Fig. 28.2).

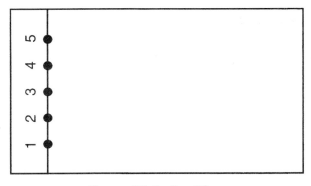

Figure 28.2 Spotting.

4. For spots # 1 and #5 use your isolated methyl esters obtained from corn oil. For spot # 2 use methyl oleate, for #3 methyl linoleate and for # 4 methyl palmitate. For each sample use a separate capillary tube. In spotting, apply each sample in the capillary to the plate until it spreads to a spot of 1 mm diameter. Dry the spots with a heat lamp. Pour about 15 mL solvent (hexane: diethyl ether; 8:2) into a 500-mL beaker. Place the spotted TLC plate diagonally for ascending chromatography. Make certain that the spots applied are **above** the surface of the eluting solvent. Cover the beaker lightly with an aluminum foil to avoid excessive solvent evaporation.

5. When the solvent front has risen to about 1 - 2 cm from the top edge, remove the plate from the beaker. Mark the advance of the solvent front with a pencil. Dry the plate with a heat lamp. Place the dried plate in a beaker containing a few iodine crystals. Cover the beaker tightly with aluminum foil. Place the beaker in a 110°C oven for 3 - 4 minutes. Remove the beaker and let it cool to room temperature. *This part is essential to avoid inhaling iodine vapors.* Remove the TLC plate from the beaker and mark the spots with a pencil.

6. Record the distance the solvent front advanced on your Report Sheet (4). Record on your Report Sheet (5) the distance of each iodine stained spot is from its origin. Calculate the R_f values of your samples (6).

CHEMICALS AND EQUIPMENT

1. Corn oil
2. Methyl palmitate
3. Methyl oleate
4. Methyl linoleate
5. Petroleum ether
6. 0.5 M KOH in ethanol
7. Concentrated HCl
8. Anhydrous Na_2SO_4
9. Methanol:perchloric acid mixture (95:5)
10. Hexane:diethyl ether mixture (8:2)
11. Iodine crystals, I_2
12. Aluminum foil
13. Polyethylene gloves
14. 15 × 6.5 cm silica gel TLC plate
16. Capillary tubes open on both ends
17. Heat lamp
18. Water bath
19. Ruler
20. Drying oven, 110°C.

EXPERIMENT 28

NAME _____ SECTION _____ DATE _____

PARTNER _____ GRADE _____

PRE-LAB QUESTIONS

1. Write the schematic structures of palmitic and oleic acids. Indicate the single bonds of the aliphatic chain with zig-zag drawing and the cis double bonds with horizontal double lines.

2. Write the formulas of the reaction converting linolenic acid to its methyl ester.

3. The R_f on a chromatogram is the ratio of the distance a particular spot traveled divided by the distance the solvent front advanced. A particular compound has an R_f value of 0.45. How far did the spot travel on a chromatogram in which the solvent front advanced 13 cm?

4. Why do you have to cool the iodine chamber (the beaker containing the chromatogram and iodine vapor) from 110°C to room temperature?

EXPERIMENT 28

NAME _____ SECTION _____ DATE _____

PARTNER _____ GRADE _____

REPORT SHEET

1. Weight of beaker _____ g
2. Weight of beaker and oil _____ g
3. Weight of oil: (2) − (1) _____ g

Distances on the chromatogram in cm

4. the solvent front _____

5. spot #1 a,b,c,d.e   a____ b____ c____ d____ e____

6. spot #2 _____

7. spot #3 _____

8. spot #4 _____

9. spot #5 a,b,c,d,e   a____ b____ c____ d____ e____

Calculated R_f values

10. for spot #1 ((5)/(4))a,b,c,d,e   a____ b____ c____ d____ e____

11. for spot #2 ((6)/(4)) _____

12. for spot #3 ((7)/(4)) _____

13. for spot #4 ((8)/(4)) _____

14. for spot #5 ((9)/(4)) a,b,c,d,e   a____ b____ c____ d____ e____

15. How many fatty acids were present in your corn oil?

16. How many fatty acids could you identify? Name the identifiable fatty acids in the corn oil.

POST-LAB QUESTIONS

1. Which of the identifiable fatty acids of your corn oil was a saturated fatty acid?

2. Judging from the iodine spots of samples 2, 3 and 4 which fatty acid reacts most strongly with iodine? Why?

3. Does the intensity of the iodine spots in the corn oil sample indicate their abundance? Explain.

4. Given two saturated fatty acids: one a short chain of 10 carbons and the other a long chain of 20 carbons, which would move faster on the TLC plate? Explain.

Analysis of Vitamin A in Margarine

BACKGROUND

Vitamin A, or retinol, is one of the major fat soluble vitamins. It is present in many foods; the best natural sources are liver, butter, margarine, egg yolk, carrots, spinach and sweet potatoes. Vitamin A is the precursor of retinal, the essential component of the visual pigment rhodopsin.

Vitamin A (All-*trans*-retinol)

11-*cis*-retinal

When a photon of light penetrates the eye, it is absorbed by the 11-*cis*-retinal. The absorption of light converts the 11-*cis*-retinal to all-*trans*-retinal:

$\xrightarrow{hv}$

11-*cis*-retinal

All-*trans*-retinal

This isomerization converts the energy of photon into an atomic motion which in turn is converted into an electrical signal. The electrical signal generated in the retina of the eye is transmitted through the optic nerve into the brain's visual cortex.

Even though part of the all-*trans*-retinal is regenerated in the dark to 11-*cis*-retinal, for good vision, especially for night vision, a constant supply of vitamin A is needed. The recommended daily allowence of vitamin A is 750 μg. Deficiency in vitamin A results in night blindness and keratinization of epithelium. The latter compromises the integrity of healthy skin. In young animals vitamin A is also required for growth. On the other hand, large doses of vitamin A, sometimes recommended in faddish diets, can be harmful. A daily dose above 1500 μg can be toxic.

OBJECTIVE

To analyze the vitamin A content of margarine by spectrophotometric method.

PROCEDURES

The analysis of vitamin A requires a multi-step process. In order that you should be able to follow the step by step procedures a flow chart is provided here:

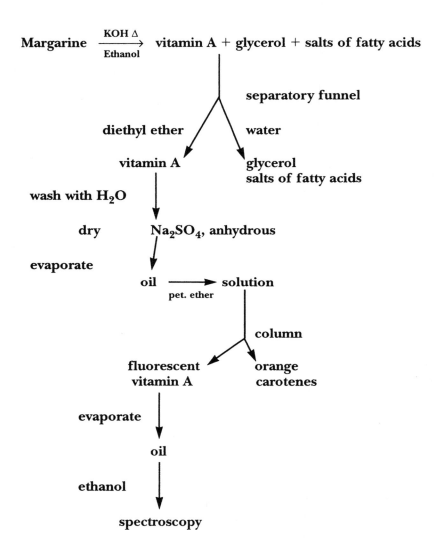

1. Margarine is largely fat. In order to separate vitamin A from the fat in margarine, first the sample must be saponified. This converts the fat to water soluble products, glycerol and potassium salts of fatty acids. Vitamin A can be extracted by diethyl ether from the products of the saponification process. To start weigh a watch glass to the nearest 0.1 g. Report this weight on the Report Sheet (1). Add approximately 10 g of margarine to the watch glass. Record the weight of watch glass plus sample to the nearest 0.1 g on your Report Sheet (2). Transfer the sample from the watch glass into a 250-mL Erlenmeyer flask with the aid of a glass rod and wash it in with 75 mL of 95% ethanol. Add 25 mL of 50% KOH solution.

(Caution: 50% KOH solution can cause burns on your skin. Handle the solution with care, do not spill it. If a drop gets on your skin wash it immediately with copious amount of water.) Cover the Erlenmeyer flask loosely with a cork and put it on an electric hot plate. Bring it gradually to a boil. Maintain the boiling for 5 min. with occasional swirling of the flask using tongs. The stirring should aid the complete dispersal of the sample. Remove the Erlenmeyer from the hot plate and let it cool to room temperature (approx. 20 min.).

2. While the sample is cooling, prepare a chromatographic column. Take a 25-mL buret. Add a small piece of glass wool. With the aid of a glass rod push it down near the stopcock. Add 15 - 16 mL petroleum ether to the buret. Open the stopcock slowly and allow the solvent to fill the tip of the buret. Close the stopcock. You should have 12 - 13 mL petroleum ether above the glass wool. Weigh about 20 g alkali aluminum oxide (alumina) in a 100-mL beaker. Place a small funnel on top of your buret. Pour the alumina slowly in small increments into the buret. Allow it to settle to form a 20 cm column. Drain the solvent but do not allow the column to run dry. **Always have at least 0.5 mL clear solvent on top of the column.** If the alumina adheres to the walls of the buret wash it down with more solvent.

3. Transfer the solution (from your reaction in step 1.) from the Erlenmeyer flask to a 500-mL separatory funnel. Rinse the flask with 30 mL distilled water and add the rinsing to the separatory funnel. Repeat the rinsing two more times. Add 100 ml diethyl ether to the separatory funnel.

Diethyl ether is very volatile and inflammable. Make it certain that there are no open flames, not even a hot electrical plate in the vicinity of the operation.

Close the separatory funnel with the glass stopper. Shake the separatory funnel vigorously (see Fig. 6.1). Allow the solution to separate into two layers. Drain the bottom aqueous layer into an Erlenmeyer flask. Add the top (diethyl ether) layer to a second clean 250-mL Erlenmeyer flask. Return the aqueous layer back into the separatory funnel. Add another 100 mL portion of diethyl ether. Shake and allow it to separate into two layers. Again drain the bottom (aqueous) layer and discard. Combine the first diethyl ether extract with the residual diethyl ether extract in the separatory funnel. Add 100 mL distilled water to the combined diethyl ether extracts in the separatory funnel. Agitate it gently and allow the water to separate. Drain and discard the washing.

4. Transfer the diethyl ether extracts into a clean 300-mL beaker. Add 3 - 5 g anhydrous Na_2SO_4 and stir it gently for 5 min. to remove traces of water. Decant the the diethyl ether extract into a clean 300-mL beaker. Add a boiling chip or a boiling stick. Concentrate the diethyl ether solution to about 25-mL volume by placing the beaker in the hood on a steam bath. Transfer the sample to a 50-mL beaker and continue to evaporate on the steam bath until an oily residue forms. Remove the beaker from the steam bath. Cool it in an ice bath for one min. Add 5 mL petroleum ether and transfer the liquid (without the boiling chip) to a 10-mL volumetric flask. Add sufficient petroleum ether to bring it to volume.

5. Add 5 mL of the extracts dissolved in the petroleum ether to your chromatographic column. By opening the stopcock drain the sample into your column but take care *not to let the column run dry.* (Always have about 0.5 mL liquid on top of the column). Continue to add

solvent to the top of your column. Collect the eluents in a beaker. First you will see the orange colored carotenes moving down the column. With the aid of an UV lamp you can also observe a fluorescent band following the carotenes. This fluorescent band contains your vitamin A. Allow all the orange color band to move to the bottom of your column and into the collecting beaker. When the fluorescent band reaches the bottom of the column, close the stopcock. Change the collecting flask to a 25-mL cylinder. Open the stopcock and by continuously adding petroleum ether to the top of the column, elute the fluorescent band from the column. Continue the elution until all the fluorescent band has been drained into the cylinder. Close the the stopcock and record the volume of the eluate in the cylinder on your Report Sheet (4). Add the vitamin A in the petroleum ether eluate to a dry and clean 50-mL beaker. Evaporate the solvent under the hood on a steam bath. The evaporation is complete when an oily residue appears in the beaker. Add 5 mL absolute ethanol to the beaker. Transfer the sample into a 10-mL volumetric flask and bring it to volume by adding absolute ethanol.

6. Place your sample in a 1 cm length quartz spectroscopic cell. The control (blank) spectroscopic cell should contain absolute ethanol. Read the absorbance of your sample against the blank, according to the instructions of your spectrophotometer, at 325 nm. Record the absorption at 325 nm on your Report Sheet (5).

7. Calculate the amount of margarine that yielded the vitamin A in the petroleum ether eluate. Remember that you added only half (5 mL) of the extract to the column. Report this value on your Report Sheet (6). Calculate the grams of margarine that would have yielded the vitamin A in 1 mL absolute ethanol by dividing (6)/10 mL. Record it on your Report Sheet (7). Calculate the vitamin A in a lb of margarine by using the following formula: μg Vitamin A/lb of margarine = Absorption $\times$ 5.5 $\times$ [454/(7)]. Record your value on the Report Sheet (8).

CHEMICALS AND EQUIPMENT

1. Separatory funnel (500-mL)
2. Buret (25-mL)
3. UV lamp
4. Spectrophotometer (near UV)
5. Margarine
6. Petroleum ether, b.p. 30 - 60°C
7. 95% ethanol
8. Absolute ethanol
9. Diethyl ether
10. Glass wool
11. Alkali aluminum oxide (alumina).

EXPERIMENT 29

NAME _____ SECTION _____ DATE _____

PARTNER _____ GRADE _____

PRE-LAB QUESTIONS

1. The structure of β-carotene is given below. What is the difference between β-carotene and vitamin A?

β-carotene

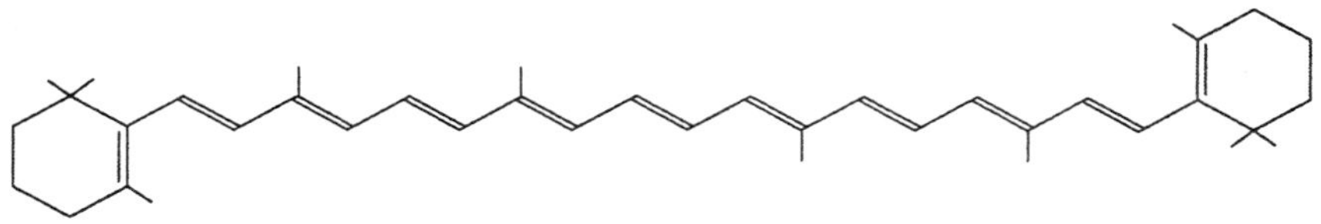

2. Why should you not have a burning Bunsen burner in the lab while you are working with diethyl ether?

3. In the saponification process you hydrolyzed fat in the presence of KOH. (See Background of Exp. 28). Write an equation of a reaction in which the fat is hydrolyzed in the presence of HCl. What is the difference between the products of the saponification and that of the acid hydrolysis?

EXPERIMENT 29

NAME _____   SECTION _____   DATE _____

PARTNER _____   GRADE _____

REPORT SHEET

1. Weight of watch glass   _____ g

2. Weight of watch glass + margarine   _____ g

3. Weight of margarine: (2) - (1)   _____ g

4. Volume of petroleum ether eluate   _____ mL

5. Absorption at 325 nm   _____

6. Grams of margarine in 1 mL of petroleum ether eluate:
    $2 \times [(3)/(4)]$ _____

7. Grams of margarine in 1 mL of absolute ethanol: (6)/10 mL   _____

8. μg vitamin A/ lb margarine:
    $(5) \times 5.5 \times [454/(7)]$ _____

POST-LAB QUESTIONS

1. Look at the structure of vitamin A; explain why is it not soluble in water.

2. In your separation scheme the fatty acids of the margarine ended up in the aqueous wash, which was discarded. Could you have removed the fatty acids similarly, if instead of saponification you used acid hydrolysis? Explain.

3. Suppose that your petroleum ether eluate had a faint orange color. Vitamin A is colorless. What could have given you the coloration?

4. The label on a commercial margarine sample states that 1 gram of it contains 15% of the daily recommended allowance. Was your sample richer or poorer in vitamin A than the above mentioned commercial sample?

TLC Separation of Amino Acids

BACKGROUND

Amino acids are the building blocks of peptides and proteins. They possess two functional groups: the carboxylic acid group gives the acidic character, and the amino group provides the basic character. The common structure of all amino acids is

$$R-\overset{\displaystyle H}{\underset{\displaystyle NH_2}{C}}-COOH$$

The R represents the side chain that is different for each of the amino acids that are commonly found in proteins. However, all 20 amino acids have a free carboxylic acid group and a free amino (primary amine) group, except proline which has a cyclic side chain and a secondary amino group.

Proline

We use the properties provided by these groups to characterize the amino acids. The common carboxylic acid and amino groups provide the acid-base nature of the amino acids. The different side chains, and the solubilities provided by these side chains can be utilized to identify the different amino acids by their rate of migration in thin layer chromatography.

In this experiment we use thin layer chromatography to identify aspartame, an artificial sweetener, and its hydrolysis products in certain foods.

Aspartame

Aspartame is the methyl ester of the dipeptide aspartylphenylalanine. Upon hydrolysis with HCl it yields aspartic acid, phenylalanine and methyl alcohol. When this artificial sweetener was approved by the Food and Drug Administration, opponents of aspartame claimed that it is a health hazard, because in soft drinks upon long storage aspartame would be hydrolyzed and would yield poisonous methyl alcohol. The Food and Drug Administration ruled, however, that aspartame is sufficiently stable and fit for human consumption. Only a warning must be put on the labels containing aspartame. This warning is for patients suffering from phenylketonurea who cannot tolerate phenylalanine.

To run a thin layer chromatography experiment we use silica gel in a thin layer on a plastic or on a glass plate. We apply the sample (aspartame or amino acids) as a spot to a strip of a thin layer plate. The plate is dipped into a mixture of solvents. The solvent moves up the thin gel by capillary action and carries the sample with it. Each amino acid may have a different migration rate depending on the solubility of the side chain in the solvent. Amino acids with similar side chains are expected to move with similar though not identical rates; those that have quite different side chains are expected to migrate with different velocities. Depending on the solvent system used, almost all amino acids and dipeptides can be separated from each other by thin layer chromatography (TLC).

We actually do not measure the rate of migration of an amino acid or a dipeptide, but rather, how far a particular amino acid travels in the thin silica gel layer relative to the migration of the solvent. This ratio is called the R_f value. In order to calculate the R_f values, one must be able to visualize the position of the amino acid or dipeptide. This is done by spraying the thin layer silica gel plate with a ninhydrin solution that reacts with the amino group of the amino acid. A purple color is produced when the gel is heated. (The proline not having a primary amine gives a yellow color with ninhydrin). For example, if the purple spot of an amino acid appears on the TLC 4.5 cm away from the origin and the solvent front migrates 9.0 cm (Fig. 30.1) the R_f value for the amino acid is

$$R_f = \frac{\text{distance traveled by the amino acid}}{\text{distance traveled by the solvent front}} = \frac{4.5 \text{ cm}}{9.0 \text{ cm}} = 0.50$$

In the present experiment you will determine the R_f values of three amino acids, phenylalanine, aspartic acid and leucine. You will also measure the R_f value of aspartame.

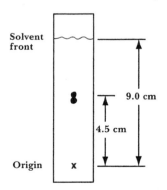

Figure 30.1 TLC chromatogram.

The aspartame you will analyze is actually a commercial sweetener, Equal by NutraSweet Co., that contains besides aspartame, silicon dioxide, glucose, cellulose and calcium phosphate. None of these other ingredients of Equal will give a purple or any other colored spot with ninhydrin. Occasionally some sweeteners may contain a small amount of leucine which can be detected by the ninhydrin test. You will also hydrolyze aspartame using HCl as a catalyst to see if the hydrolysis products will prove that the sweetener was truly aspartame. Finally you will analyze some commercial soft drinks, supplied by your instructor.

The analysis of the soft drink can tell you if the aspartame was hydrolyzed at all during the processing and storing of the soft drink.

OBJECTIVES

1. To separate amino acids and a dipeptide by TLC.
2. To identify hydrolysis products of aspartame.
3. To analyze the state of aspartame in soft drinks.

PROCEDURE

1. Dissolve about 10 mg of the sweetener Equal in 1 mL of 3 M HCl in a test tube. Heat it with a Bunsen burner to boil for 30 sec., but make sure the liquid should not completely evaporate. Cool the test tube and label it as "Hydrolyzed Aspartame".

2. Label 5 additional small test tubes, respectively, for aspartic acid, phenylalanine, leucine, aspartame and Diet Coca-Cola. Place about 0.5 mL samples in each test tube.

3. Take two 15 × 6.5 cm TLC plates. With a pencil lightly draw a line parallel to the 6.5 cm edge and about 1 cm from the edge. Mark the positions of 5 spots on each plate, spaced equally, where you will spot your samples (Fig. 30.2). You must make sure that you don't touch the plates with your fingers.

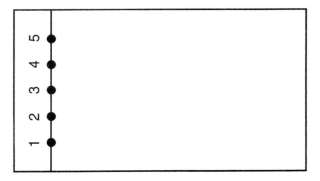

Figure 30.2 Spotting.

Either use plastic gloves or handle the plates by holding them only at their edges. This precaution must be observed throughout the whole operation, because amino acids from your fingers will contaminate the plate.

On plate A you will spot samples of (1) phenylalanine, (2) aspartic acid, (3) leucine, (4) aspartame in Equal and (5) hydrolized aspartame you prepared in step 1. On plate B you will spot samples of Diet Coca-Cola on lanes (1) and (4), aspartic acid on lane (2), aspartame in Equal on lane (3), and the hydrolyzed aspartame you prepared previously on lane (5).

4. First spot plate A. For each sample use a separate capillary tube. Apply the sample to the plate until it spreads to a spot of 1 mm diameter. Dry the spots. (If a heat lamp is available, use it for drying). Pour about 15 mL of solvent mixture (butanol:acetic acid:water) into a large (500-mL or 1-L) beaker and place your spotted plate diagonally for ascending chromatography. Make certain that the spots applied to the plate are above the surface of the eluting solvent. Cover the beaker with aluminum foil to avoid the evaporation of the solvent mixture.

5. Spot plate B. For aspartic acid, lane (2), and for the hydrolyzed and non-hydrolyzed aspartame, lanes (3) and (5), use one spot as before. For Coca-Cola [lanes (1) and (4)] multiple spotting is needed. Apply the capillary tube 12 - 15 times to the same spot, making certain that between each application the previous sample has been dried. Also try to control the size of the spots that they should not spread too much, not more than 2 mm in diameter. Dry the spots as before. Place the plate in a large beaker containing the eluting solvent as before. Cover the beaker with aluminum foil. Allow about 50 - 60 minutes for the solvent front to advance.

6. When the solvent front nears the edge of the plate, about 1 - 2 cm from the edge, remove the plate from the beaker. You must not allow the solvent front to advance up to or beyond the edge of the plate. Mark immediately *with a pencil* the position of the solvent front. Under a hood dry the plates with the aid of a heat lamp or hair dryer. Use polyethylene gloves when spraying the dry plates with the ninhydrin solution. *Be careful not to spray ninhydrin on your hand and not to touch the sprayed areas with bare hands. If the ninhydrin spray touches your skin which contains amino acids, your fingers will be discolored for a few days.* Place the sprayed plates into a drying oven at 105 - 110°C for 2 - 3 minutes.

7. Remove the plates from the oven. Mark the center of the spots and calculate the R_f values of each spot.

CHEMICALS AND EQUIPMENT

1.   0.1% solutions of aspartic acid, phenylalanine and leucine
2.   0.5% solution of aspartame (Equal)
3.   Diet Coca-Cola
4.   3 M HCl
5.   0.2% ninhydrine spray
6.   Butanol: acetic acid: water solvent mixture
7.   Equal sweetener
8.   Aluminum foil
9.   15 × 6.5 cm silica gel TLC plates
10. Ruler
11. Polyethylene gloves
12. Capillary tubes open on both ends
13. Heat lamp or hair dryer
14. Drying oven, 110°C

EXPERIMENT 30

NAME _____ SECTION _____ DATE _____

PARTNER _____ GRADE _____

PRE-LAB QUESTIONS

1. If an amino acid moved 2.1 cm on a TLC plate and the solvent moved 7.0 cm, what is the R_f value of the amino acid?

2. Why must you use a pencil and not ink to mark the origin of your spot on the TLC plate?

3. What would happen if you didn't use gloves and your finger comes in contact with the ninhydrin spray?

4. How would you calculate the R_f values of your samples if you would allow the solvent front to run over the edges of your TLC plates?

EXPERIMENT 30

NAME _____ SECTION _____ DATE _____

PARTNER _____ GRADE _____

REPORT SHEET

1. **Sample**   **Distance traveled (mm)**   **Solvent front (mm)** $\underline{R_f}$

   Aspartic acid

   Phenylalanine

   Leucine

   Aspartame

   Hydrolyzed

   aspartame

   Diet Coca-Cola

2. Identification

   (a) Name the amino acids you found in the hydrolysate of the sweetener, Equal.

   (b) How many spots were stained with ninhydrin (1) in Equal and (2) in Coca-Cola samples?

POST-LAB QUESTIONS

1. Can you separate aspartame, a dipeptide, from its constituent amino acids by TLC technique? Which sample migrated the slowest?

2. In testing the hydrolysate of aspartame, you forgot to mark the position of the solvent front on your TLC plate. Could you:

   (a) determine how many amino acids were in the aspartame;

   (b) identify those amino acids?

3. Do you have any evidence that the aspartame was hydrolyzed during the processing and storage of the Diet Coca-Cola sample? Explain.

4. How could you have achieved a better separation on your TLC plate between aspartic acid and phenylalanine?

5. If the R_f values of two amino acids are 0.3 and 0.35, respectively, what length should the TLC plate be in order that the two spots should be separated by 1 cm?

Isolation and Identification of Casein

EXPERIMENT 31

BACKGROUND

Casein is the most important protein in milk. It functions as a storage protein fulfilling nutritional requirements. Casein can be isolated from milk by acidification to bring it to its iso-electric point. At the isoelectric point the number of positive charges on a protein equal the number of negative charges. Proteins are least soluble in water at their isoelectric points because they tend to aggregate by electrostatic interaction. The positive end of one protein molecule attracts the negative end of another protein molecule and the aggregates precipitate out of solution.

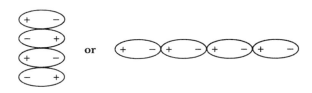

On the other hand, if a protein molecule has a net positive (at low pH or acidic condition) or a net negative charge (at high pH or basic condition), its solubility in water is increased.

$$\overset{+}{N}H_3 \sim COOH \xleftarrow[\text{low pH}]{H^+} \overset{+}{N}H_3 \sim\sim COO^- \xrightarrow[\text{high pH}]{OH^+} NH_2 \sim COO^- + H_2O$$

More soluble **Least soluble** **More soluble**

(At isoelectric pH)

In the first part of this experiment you are going to isolate casein from milk which has a pH of about 7. Casein will be separated as an insoluble precipitate by acidification of the milk to its isoelectric point (pH = 4.6). The fat that precipitates along with casein can be removed by dissolving it in alcohol.

In the second part of this experiment you are going to prove that the precipitated milk product is a protein. The identification will be achieved by performing a few important chemical tests.

287

1. **The Biuret Test.** This is one of the most general tests for proteins. When a protein reacts with copper(II) sulfate, a positive test is the formation of a copper complex which has a violet color.

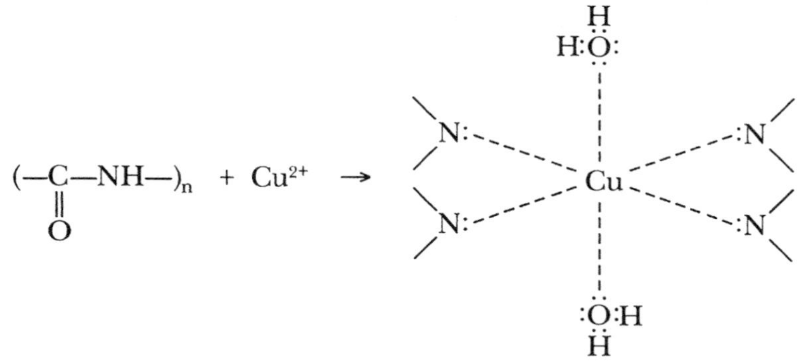

| Protein | Blue color | Protein-copper complex (Violet color) |

This test works for any protein or compound that contains two or more of the following groups:

$$-\underset{\underset{O}{\|}}{C}-NH-, \quad -\underset{\underset{O}{\|}}{C}-NH_2, \quad -CH_2-NH_2, \quad -\underset{\underset{NH}{\|}}{C}-NH_2, \quad -\underset{\underset{S}{\|}}{C}-NH_2$$

2. **The Ninhydrin Test.** Amino acids with a free $-NH_2$ group and proteins containing free amino groups react with ninhydrin to give a purple-blue complex.

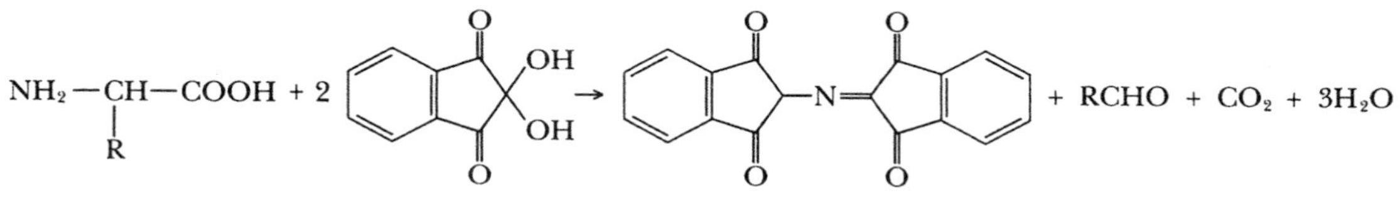

| Protein | Ninhydrin | Purple-blue complex |

3. **Heavy Metal Ions Test.** Heavy metal ions precipitate proteins from their solutions. The ions that are most commonly used for protein precipitation are Zn^{2+}, Fe^{3+}, Cu^{2+}, Sb^{3+}, Ag^+, Cd^{2+}, and Pb^{2+}. Among these metal ions, Hg^{2+}, Cd^{2+}, and Pb^{2+} are known for their notorious toxicity. They can cause serious damage to proteins (especially the enzymes) by denaturing them. This can result in death. The precipitation occurs because proteins become cross-linked by heavy metals as shown below:

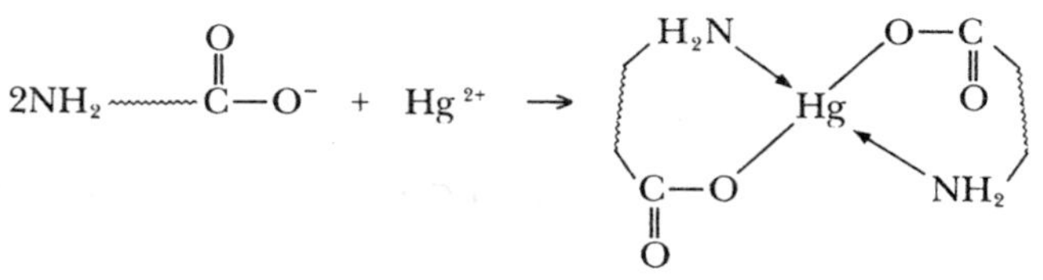

Insoluble precipitate

Victims swallowing Hg^{2+} or Pb^{2+} ions are often treated with an antidote of a food rich in proteins which can combine with mercury or lead ions in the victim's stomach and, hopefully, prevent absorption! Milk and raw egg white are used most often. The insoluble complexes are then immediately removed from the stomach by an emetic.

4. **The Xanthoprotein Test.** This is a characteristic reaction of proteins that contain phenyl rings:

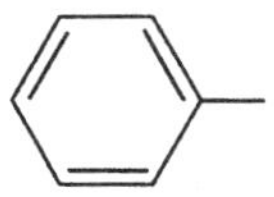

Concentrated nitric acid reacts with the phenyl ring to give a yellow-colored aromatic nitro compound. Addition of alkali at this point will deepen the color to orange.

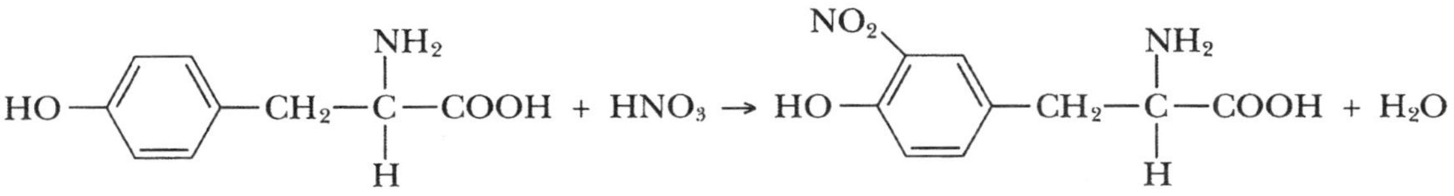

$$HO-\hspace{-0.3em}\bigcirc\hspace{-0.3em}-CH_2-\underset{\underset{H}{|}}{\overset{\overset{NH_2}{|}}{C}}-COOH \;+\; HNO_3 \;\rightarrow\; HO-\hspace{-0.3em}\bigcirc\hspace{-0.3em}-CH_2-\underset{\underset{H}{|}}{\overset{\overset{NH_2}{|}}{C}}-COOH \;+\; H_2O$$

                    Tyrosine   **Colored compound**

The yellow stains on the skin caused by nitric acid are the result of the xanthoprotein reaction.

OBJECTIVES

1. To isolate the casein from milk under isoelectric conditions.
2. To perform some chemical tests to identify proteins.

PROCEDURE

A. *Isolation of Casein*

1. To a 250-mL Erlenmeyer flask, add 50.00 g of milk and heat the flask in a water bath (a 600-mL beaker containing about 200 mL of tap water; see Fig. 31.1). Stir the solution constantly with a stirring rod. When the bath temperature has reached about 40°C, remove the flask from the water bath and add about 10 drops of glacial acetic acid while stirring. Observe the formation of a precipitate.

2. Filter the mixture into a 100-mL beaker by pouring it through a cheese cloth which is fastened with a rubber band over the mouth of the beaker (Fig. 31.2). Remove most of the water from the precipitate by squeezing the cloth gently. Discard the filtrate in the beaker. Using a spatula, scrape the precipitate from the cheese cloth into the empty flask.

3. Add 25 mL of 95% ethanol to the flask. After stirring the mixture for 5 min., allow the solid to settle. Carefully decant (pour off) the liquid that contains fats into a beaker. Discard the liquid.

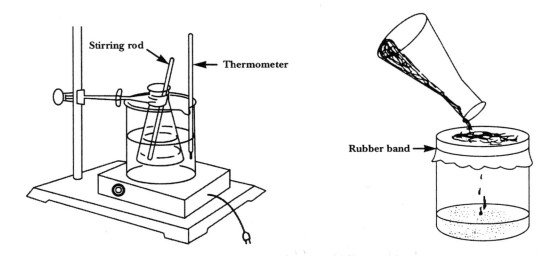

Figure 31.1 Precipitation of casein.   **Figure 31.2 Filtration of casein.**

4. To the residue, add 25 mL of 1:1 mixture of diether-ethanol. After stirring the resulting mixture for 5 min, collect the solid by vacuum filtration.

5. Spread the casein on a paper towel and let it dry. Weigh the dried casein and calculate the percentage of casein in the milk.

$$\% \text{ casein} = \frac{\text{weight of solid (casein)}}{50.00 \text{ g of milk}} \times 100$$

B. Chemical Analysis of Proteins

1. **Biuret Test.** Place 15 drops of each of the following solutions in 5 clean, labeled test tubes.
   a. 2% glycine
   b. 2% gelatin
   c. 2% albumin
   d. Casein prepared in part A (one quarter of a full spatula) + 15 drops of distilled water
   e. 1% tyrosine

   To each of the test tubes, add 5 drops of 10% NaOH solution and two drops of a dilute $CuSO_4$ solution while swirling. The development of purplish violet color is evidence of the presence of proteins. Record your results on the Report Sheet.

2. **The Ninhydrin Test.** Place 15 drops of each of the following solutions in 5 clean, labeled test tubes.
   a. 2% glycine
   b. 2% gelatin
   c. 2% albumin
   d. Casein prepared in part A (one quarter of a full spatula) + 15 drops of distilled water
   e. 1% tyrosine

   To each of the test tubes, add 5 drops of ninhydrin reagent and heat the test tubes in a boiling water bath for about 5 min. Record your results on the Report Sheet.

EXPERIMENT

32

Isolation and Identification of DNA from Yeast

BACKGROUND

Hereditary traits are transmitted by genes. Genes are parts of giant deoxyribonucleic acid (DNA) molecules. In lower organisms, such as bacteria and yeast, both DNA and RNA (ribonucleic acid) occur in the cytoplasm while in higher organisms, most of the DNA is inside the nucleus and the RNA is outside the nucleus, in other organelles and in the cytoplasm.

In this experiment we will isolate DNA molecules from yeast cells. The first task is to break up the cells. This is achieved by a combination of different techniques and agents. Grinding up the cells with sand disrupts them and the cytoplasm of many yeast cells is spilled out. However, this is not a complete process. The addition of a detergent, hexadecyltrimethylammonium bromide, CTAB, accomplishes two functions: 1) it helps to solubilize cell membranes and and thereby further weakens the cell structure and 2) it helps to inactivate the nucleic acid degrading enzymes, nucleases, that are present. The addition of a chelating agent, ethylenediamine tetraacetate, EDTA, also inactivates these enzymes. EDTA removes the di- and tri-valent cations which are necessary for the activity of nucleases. Without this inhibition the nucleases would degrade the nucleic acids to their constituent nucleotides. The final assault on the yeast cell is the osmotic shock. This is provided by a hypotonic saline-EDTA solution. The already weakened cells (by grinding and treatment with CTAB) will burst in the hypotonic medium and spill their contents, nucleic acids among them.

Once the nucleic acids are in solution they must be separated from the other constituents of the cell. First the protein molecules must be removed. Many of the proteins of the cell are strongly associated with nucleic acids. The addition of sodium perchlorate ($NaClO_4$) dissociates the proteins from nucleic acids. When the mixture is shaken with the organic solvent (chloroform-isoamyl alcohol) the proteins are denatured and they precipitate at the interface. At the same time the lipid components of the cells are dissolved in the organic solvent. Thus the aqueous layer will contain nucleic acids, small water soluble molecules and even some proteins as contaminants.

The addition of ethanol precipitates the large molecules (DNA, RNA and proteins) and leaves the small molecules in solution. DNA, being the largest fibrous molecule, forms thread-like precipitates that can be spooled off on a rod. The protein and RNA form a gelatinous precipitate that cannot be picked up by winding them on a glass rod. Thus the spooling separates DNA from RNA.

After the isolation of DNA we will probe its identity by the diphenylamine test. The blue color of this test is specific for deoxyribose and the appearance of blue color can be used to identify the deoxyribose containing DNA molecule.

FLOW DIAGRAM OF THE DNA ISOLATION PROCESS.

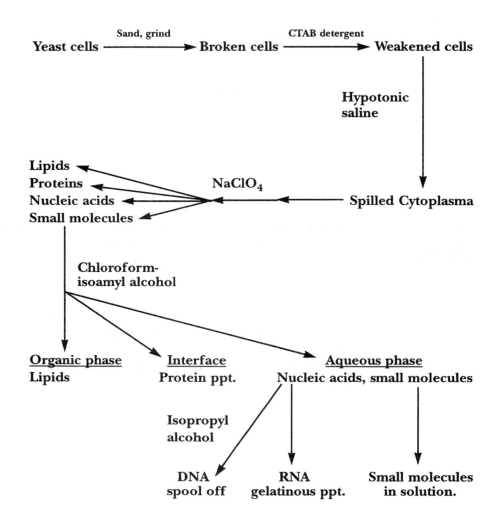

OBJECTIVES

To demonstrate the separation of DNA molecules from other cell constituents and to prove their identity.

PROCEDURE

1. Cool a mortar in ice water. Add 2 to 3 g of baker's yeast and twice as much acid washed sand. Grind the yeast and the sand vigorously with a pestle for 5 - 10 minutes to disrupt the cells. (Two groups can work together in grinding and then divide the product.)

2. Preheat 25 mL of CTAB isolation buffer (2% CTAB, 0.15 M NaCl, 0.2% 2-mercaptoethanol, 20 mM EDTA and 100 mM Tris-HCl at pH 8.0) in a 100-mL beaker in a 60°C water bath.

3. Add the ground yeast and sand to the saline-CTAB solution. Mix the solution with the sand. Let it stand for 20 min., with occasional swirling, all the while maintaining the temperature at 60°C.

4. Decant the cell suspension into a 250-mL Erlenmayer flask leaving the sand behind. Cool the solution to room temperature. Add 5 mL of 6 M $NaClO_4$ solution and mix well. Transfer 40 mL chloroform-isoamyl alcohol mixture into the flask. Stopper the flask with a cork. Shake it for 10 minutes, sloshing the contents from side to side once every 15 sec. A frothy emulsion will form. After 10 min. let the emulsion settle.

5. Break up the emulsion by gently swirling with a glass rod reaching into the interface. The complete separation into two distinct layers is not possible without centrifugation. (If desk top centrifuges are available it is preferable to separate the layers by centrifuging at 1600 × g for 5 min.) However, one can proceed without centrifugation as well. When sufficient amount (20 - 30 mL) of the top aqueous layer is cleared, remove this with a Pasteur pipet and transfer it to a graduated cylinder. Measure the volume and pour the contents into a 250-mL beaker. Pay attention that none of the brownish precipitate, droplets of emulsion, is transferred.

6. To the viscous DNA containing aqueous solution, add slowly, twice its volume of cold isopropyl alcohol, taking care that the alcohol flows along the side of the beaker settling on top of the aqueous solution. With a flame sterilized glass rod gently stir the DNA-isopropyl alcohol solution. This procedure is *critical*. The DNA will form a thread-like precipitate. *Rotating (not stirring) the glass rod* spools all the DNA precipitate onto the glass rod. As the DNA is wound on the rod squeeze out the excess liquid by pressing the rod against the wall of the beaker. Transfer the spooled DNA on the rod into a test tube containing 95% ethanol.

7. Discard the alcohol solution left in the beaker and the chloroform-isoamyl alcohol solution left in the Erlenmayer flask into specially labelled waste jars. Do not pour them down the sink.

8. Remove the rod and the spooled DNA from the test tube. Dry the DNA with a clean filter paper. Note its appearance. Dissolve the isolated crude DNA in 2 mL citrate buffer (0.15 M NaCl, 0.015 M sodium citrate). Set up 4 dry and clean test tubes. Into the test tubes add 2 mL each of the following:

Test tube	Solution
1	1% glucose
2	1% ribose
3	1% deoxyribose
4	crude DNA solution

Add 5 mL diphenylamine reagent to each test tube.

CAUTION!   *Diphenylamine reagent contains glacial acetic acid and concentrated sulfuric acid. Handle with care. Use gloves.*

Mix the contents of the test tubes. Heat the test tubes in boiling water bath for 10 min. Note the color.

CHEMICALS AND EQUIPMENT

1. Baker's yeast
2. Sand
3. Saline-CTAB isolation buffer
4. $NaClO_4$ solution
5. Chloroform-isoamyl alcohol solvent
6. Citrate buffer
7. Isopropyl alcohol
8. Glucose solution
9. Ribose solution
10. Deoxyribose solution
11. Diphenylamine reagent
12. 95% ethanol
13. Mortar and pestle.
14. Desk top clinical centrifuges (optional)

EXPERIMENT 32

NAME _____ SECTION _____ DATE _____

PARTNER _____ GRADE _____

PRE-LAB QUESTIONS

Consult the Chapter on DNA of your text book to answer the structural questions.

1. Draw the structure of the purine and pyrimidine bases that are part of the DNA molecule.

2. Draw the structures of ribose and deoxyribose.

3. Why must you handle the diphenylamine reagent with great care?

4. The most demanding part of this experiment is grinding the yeast cells with sand. What does this process accomplish? Would you be able to isolate DNA without this grinding?

EXPERIMENT 32

NAME _____ **SECTION** _____ **DATE** _____

PARTNER _____ **GRADE** _____

REPORT SHEET

1. Describe the appearance of the crude DNA preparation.

2. Diphenylamine test.

<u>Solution</u>	<u>Color</u>
1% glucose	_____
1% ribose	_____
1% deoxyribose	_____
crude DNA sample	_____

    Did the diphenylamine test confirm the identity of DNA?

3. How did you separate the DNA from proteins?

4. After mixing the aqueous extract with chloroform-isoamyl alcohol mixture, which layer contained the RNA (aqueous or organic)?

5. What compounds were left behind in the isopropyl alcohol solution after spooling the DNA?

POST-LAB QUESTIONS

1. Can the diphenylamine reagent distinguish between ribose and deoxyribose and between DNA and RNA?

2. If you forgot to add $NaClO_4$ solution to your extract how would the omission effect the purity of your DNA preparation?

3. Why can we isolate DNA from the precipitate which also contains RNA and proteins by the simple "spooling" procedure?

Infrared Spectroscopy

Appendix

Once a compound has been isolated from a natural source or synthesized in the laboratory, the task of the chemist is to characterize the material. Melting points, boiling points and qualitative tests for functional groups (refer to the tests for aldehydes and ketones in Expt. 15 as examples) enable the chemist to compare the results to values found in the literature. Correspondence of the physical properties and chemical tests to information found in the literature serve as a means of identification.

Infrared radiation spectroscopy (IR) is another tool that a chemist can use for a structural analysis. Although largely supplemented by nuclear magnetic resonance spectroscopy (NMR), nevertheless, IR is a quick and effective technique for determining the presence of functional groups. Knowledge of the principal functional group in an organic compound goes a long way towards a complete structure determination. The analysis can be carried out whether the material is a solid, a liquid or a gas. In addition, a relatively small amount of material is used in infrared spectroscopy, thus, loss of material is minimized.

The energy associated with the absorption of infrared radiation is related to the vibrations of atoms within molecules. In organic molecules the bonds seem to bend and stretch as the atoms they connect move from the average value of their interatomic distance. It is as if the atoms were connected by a flexible spring (the covalent bond) and the springs are stretched or bent. Fig. I.1 illustrates some fundamental modes of vibration.

The energy levels associated with vibrational motions are **quantized**. This means that in order for a molecule to move from one vibrational level to another, higher one, that is to become "excited", infrared radiation of a specific wavelength is required. When this occurs energy is absorbed and the atomic vibrations become more vigorous. The particular frequency or wavelength of absorption depends on a number of factors. These factors include: the relative atomic masses of the atoms connected in a bond; the strength of the bond; the geometry of the atoms undergoing vibration (Fig. I.1).

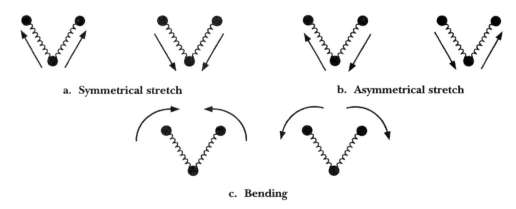

a. **Symmetrical stretch**   b. **Asymmetrical stretch**

c. **Bending**

Figure I.1 In-plane stretching and bending for a hypothetical 3-atom system.

What makes IR spectroscopy useful to the practicing organic chemist is that the vibrational frequency of a specific bond shows only small degrees of variation even when other atoms and bonds are nearby. This enables assignment of characteristic frequencies for bending or stretching modes to specific atomic groupings (functional groups).

The IR spectrum is determined by an infrared spectrophotometer which operates over an effective infrared range of 2 to 25 μm, the wavelength (λ) expressed in micrometers. A more useful expression is the wavenumber ($\bar{\nu}$), the reciprocal of the wavelength in centimeters (cm^{-1}).

$$\bar{\nu} = \frac{1}{\lambda \text{ (cm)}} = \frac{10^4}{\lambda (\mu\text{m})}$$

Most organic compounds have absorption bands extending from 4000 cm^{-1} to 400 cm^{-1}.

$$\bar{\nu} = \frac{10^4}{2.5 \ \mu\text{m}} = 4000 \text{ cm}^{-1} \ ; \quad \bar{\nu} = \frac{10^4}{25 \ \mu\text{m}} = 400 \text{ cm}^{-1}$$

In terms of energy, higher energy absorptions correspond to higher wavenumbers.

Two regions of the IR spectrum are useful for identification purposes. For the purpose of *functional group identification*, the region from 4000 cm^{-1} to 1600 cm^{-1} is most useful. The region from 1600 cm^{-1} to 400 cm^{-1} is often referred to as the *fingerprint region*. Absorption bands in this region come from more complex vibrational motions in the molecule and tend to be unique for a given compound. Table I.1 lists characteristic frequencies with the functional groups responsible for the absorptions.

Since instruments vary in the degree of sophistication, none will be described here. However, what the instrument does is to register the position of all absorption bands in the infrared region and plot them on calibrated paper. It also measures the intensity of each of the absorption bands. In general, absorption intensity (measured as % transmission of light) is plotted on the y-axis, and the wavelength or wavenumber is plotted on the x-axis; this is the infrared spectrum. Fig. I.2 correlates the principal functional groups with their relative positions on the IR chart.

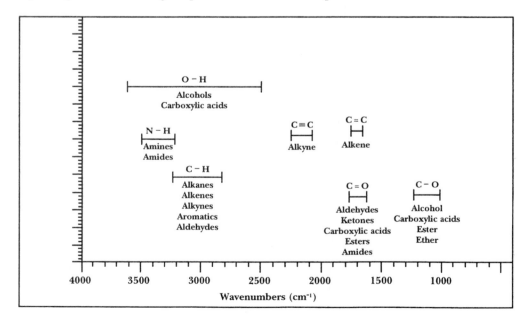

Figure I.2 Correlation chart of principal functional groups.

TABLE I.1 INFRARED ABSORPTIONS FOR SELECTED ORGANIC FUNCTIONAL GROUPS

<u>NAME</u>	<u>Functional Group</u>	<u>Absorption Range $(\bar{\nu})$ cm^{-1}</u>	<u>Type (intensity)</u>*
Alkane	C−H	2960 - 2850	S (s)
	CH$_2$	1490 - 1450	B (m)
	CH$_3$	1425 - 1375	B (m)
Alkene	C=C−H	3100 - 3000	S (m)
	C=C	1680 - 1620	S (w)
Alkyne	C≡C−H	3300 - 3200	S (s)
	C≡C	2260 - 2100	S (m)
Aromatic	Ar−H	3100 - 3000	S (m)
	ring C ≡ C	2000 - 1660	C (m)
	ring C ≡ C	1600 - 1450	S (m)
Alcohol	O−H (free)	3640 - 3610	S (m)
(Phenol)	O−H (H-bonded)	3400 - 3200	S (s,b)
	C−O	1150 - 1050	S (s)
Aldehyde	C−H (two bands)	2830 - 2720	S (s)
	C=O (alkyl)	1740 - 1720	S (s)
	C=O (conjugated)	1715 - 1695	S (s)
Ketone	C=O (alkyl)	1725 - 1705	S (s)
	C=O (conjugated)	1700 - 1660	S (s)
Carboxylic	O−H (H-bonded)	3300 - 2500	S (w,b)
acid	C=O (alkyl)	1725 - 1700	S (s)
	C=O (conjugated)	1700 - 1680	S (s)
	C−O	1250 - 1240	S (s)
Ester	C=O (alkyl)	1750 - 1730	S (s)
	C=O (conjugated)	1740 - 1725	S (s)
	C−O	1250 - 1050	S (m)
Amide	N−H	3500 - 3300	S (s)
	C=O	1700 - 1640	S (s)
	C−N	1450 - 1400	S (m)
Amine	N−H (two bands for NH$_2$)	3600 - 3200	S (s)
	N−H	1680 - 1580	B (m)
	C−N	1250 - 1050	S (m)
Ether	C−O	1150 - 1070	S (s)
Nitrile	C≡N	2260 - 2210	S (m)
Alkyl halide	C−Cl	850 - 800	S (s)
	C−Br	680 - 500	S (s)

*   S = stretch, B = bend, C = combination
    s = strong, m = medium, w = weak, b = broad

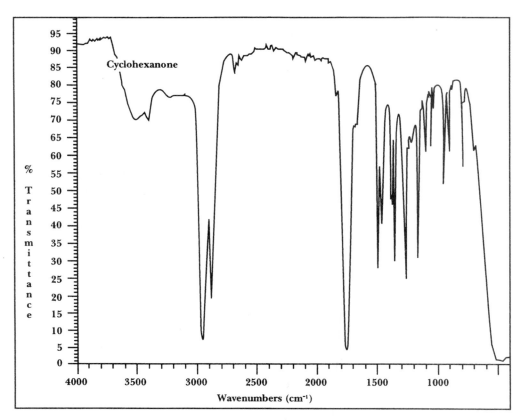

Figure I.3 Infrared spectrum of cyclohexanone, neat.

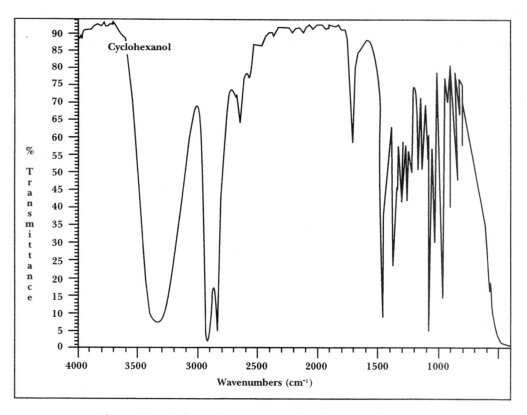

Figure I.4 Infrared spectrum of cyclohexanol, neat.

An example of an infrared spectrum is shown for cyclohexanone (Fig. I.3). Prominent absorption bands are: strong C−H stretches at 2945 and 2860 cm^{-1}; a strong C=O stretch at 1719 cm^{-1}. Numerous absorption bands appear below 1500 cm^{-1} (e.g. a medium CH$_2$ bend at 1455 cm^{-1}); these are bands characteristic of the "fingerprint" of cyclohexanone. Compare this spectrum to the IR spectrum of cyclohexanol (Fig. I.4). Prominent absorption bands are: a broad stretch at 3345 cm^{-1}, characteristic of a hydrogen bonded O−H group; strong C−H stretches at 2936 and 2860 cm^{-1}. The fingerprint region is clearly different; of importance in this region is the CH$_2$ bend at 1455 cm^{-1} (m) and the C−O stretch at 1072 cm^{-1} (m). (Note the carbonyl impurity at 1719 cm^{-1}.) An experienced spectroscopist can analyze and interpret all the absorption patterns and gain significant information about chemical structure. However, our purpose here is to use infrared spectroscopy only as an aid in the identification of organic compounds.

In general, for identification purposes, there are several goals to keep in mind:

1. look for absorption bands characteristic of particular functional groups;
2. the absence of particular absorption band(s) often is as significant as the presence of a band(s); for example, the inclusion of an impurity may cause a spurious band to appear;
3. two pure samples that give the same IR spectrum, particularly in the fingerprint region, peak for peak, must be the same compound;
4. two pure samples that do not have matching spectra must be different compounds.

IR analysis of an organic compound can be carried out by four common methods:

1. as a thin liquid film between IR discs or windows prepared from salts such as NaCl, KBr and AgCl; this is referred to as a "neat" solution; these salts are used since they do not absorb in the infrared region of the spectrum;
2. as solutions of a solid or a liquid in a solvent such as CCl$_4$ or CS$_2$; the solution is placed in a cell composed of salts as in (1) above; the solvents used have absorbtion bands in regions of the infrared spectrum that do not interfere with most organic compounds;
3. as a dispersion of a solid sample in Nujol (mineral oil) or a perfluoro-kerosene; the dispersion is then placed between plates, as in (1) above; compensation has to be made for absorptions of the dispersing agent that can interfere with the absorptions of the organic compound;
4. as a solid dispersion in KBr powder, which is then pressed by high pressure into a wafer-thin pellet.

Choice of a method depends on the sample and on the convenience of the method.

Sample preparation for an infrared spectrum of a liquid.

1. Carefully remove the sodium chloride discs from the storage desiccator. Hold the discs by the edges. Do not place your fingers on the flat faces of the discs, since moisture from your fingers will damage them. Place them on a clean tissue or Kimwipe.

2. With a Pasteur pipet, deposit one drop of the liquid sample on one of the discs.

3. Quickly place the other disc on the one containing the sample.

4. Carefully place both discs into the sample holder (Fig I.5) and tighten the assembly using only finger pressure.

5. Place the assembly into the spectrophotometer and measure the spectrum.

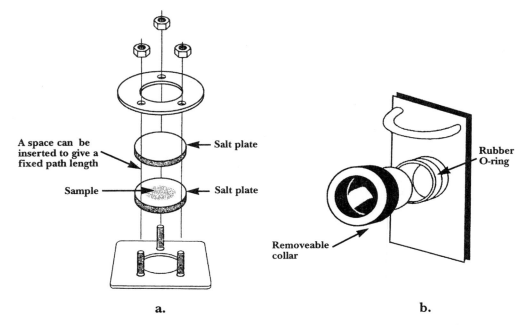

Figure I.5 Assemblies for a liquid sample: a) demountable; b) Presslok.

6. Remove the assembly from the spectrophotometer as soon as the spectrum has been completed.

7. Disassemble the sample holder and wash the sodium chloride discs with methylene chloride, in the hood, using a Pasteur pipet. Wipe the discs with a clean Kimwipe or tissue.

8. Return the discs to the desiccator.

Sample preparation for an infrared spectrum of a solid.

Solid samples can be obtained by preparing a solution, a mull or a KBr pellet. The KBr method will be discussed here since there is no interference from suspending agents (CCl_4, CS_2, Nujol); also, KBr is transparent in the infrared range.

1. Carefully grind 3 - 4 mg of sample with 70 mg of anhydrous spectral grade KBr. What we want to make is a 5% solid solution of sample in KBr. Grind with an agate mortar and pestle for 3 - 4 min. or until you get a fine, homogenous powder.

2. The most common press for student use is a KBr-Minipress and is illustrated in Fig. I.6a. Screw on one of the polished bolts and hold it with a vise or a crescent wrench.

3. Pour the ground mixture into the open-end of the minipress body and tap the body to make a uniform layer (Fig. I.6b).

4. Screw the second bolt into the minipress and turn it tightly with another wrench (Fig. I.6c). The bolts must be turned very tightly, but be careful not to shear off the heads!

5. Allow approx. 30 sec. for the powder to set. The object is to obtain a very thin, translucent wafer.

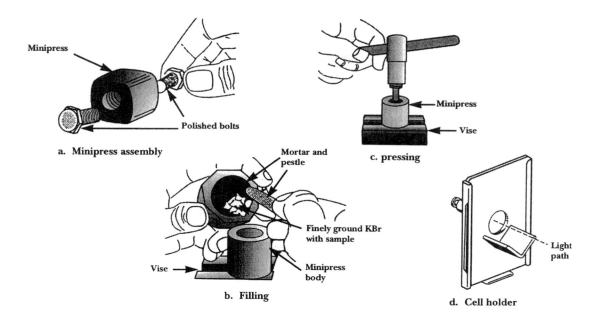

Figure I.6   Sample preparation and assembly for a KBr pellet.

6. Carefully loosen the bolts, and unscrew them from the minipress body. Be careful not to dislodge the pellet that is pressed in the body of the minipress. If the pellet is properly prepared, it will stay in the grooves of the minipress; if it cracks or collapses, the procedure must be repeated. You may need to modify the ratio of sample to KBr powder until a good pellet is obtained.

7. Place the minipress body on the cell holder (Fig. I.6d) so that the beam from the spectrophotometer passes through the KBr pellet. Record the spectrum.

8. Dislodge the pellet from the minipress with a microspatula. Then wash with water to remove the remaining solid. Rinse with acetone and dry in an oven.

9. Return the minipress and KBr powder to the storage desiccator.

Nuclear Magnetic Resonance Spectroscopy

Infrared spectroscopy is a useful technique for functional group determination. However, knowledge of the functional group in an organic molecule is only part of the answer when working out a molecular structure. Since most of an organic molecule is composed of carbon and hydrogen, learning how the hydrogens are arranged along the carbon backbone becomes an important goal in a total structure determination. Nuclear magnetic resonance spectroscopy (NMR) is a method that enables the chemist to gain this information.

The theory behind NMR is relatively simple. The nuclei of some kinds of atoms act like tiny spinning bar magnets. In the absence of an external magnetic field, the magnetic fields of the spinning nuclei are randomly oriented in space. However, when placed in a strong external magnetic field, the nuclei become lined up. For example, protons can align themselves in two arrays: parallel to the applied field or against it. In NMR, the energy required to change the alignment of the nuclei is measured.

Nuclei of certain isotopes possess a mechanical spin, or angular momentum. The total angular momentum depends on the nuclear spin, or spin number I, which may have values of 0, 1/2, 1, 3/2, ... (depending on the particular nucleus). The numerical value of the spin number I is related to the mass number and the atomic number as follows:

Mass no.	Atomic no.	Spin no., I
odd	even or odd	1/2, 3/2, 5/2, ...
even	even	0
even	odd	1, 2, 3, ...

Atoms found in organic molecules and frequently used in NMR spectroscopy belong to the first group. They include hydrogen (^1H), fluorine (^{19}F), phosphorus (^{31}P) and the carbon isotope, carbon-13 (^{13}C). The discussion which follows will focus on the hydrogen nucleus, the proton.

Since an electric charge (positive) is associated with an atomic nucleus, the spinning nucleus generates a magnetic field whose axis coincides with the axis of the spin. The nucleus is equivalent to a minute magnet of magnetic moment μ. Each nucleus for which $I > 0$ will have a characteristic magnetic moment. If a magnetic nucleus is placed in a uniform magnetic field, it is found that the magnetic dipole assumes only a discrete set of orientations. The system is said to be *quantized*.

The magnetic nucleus may assume any of $2I + 1$ orientations with respect to the direction of the applied magnetic field. Thus, quantum mechanics show that the proton (with $I = 1/2$) may have two possible orientations.

$$2I + 1 = (2)(1/2) + 1 = 2 \tag{1}$$

The magnetic moment, μ, is proportional to the spin number, I: $\mu \sim I(I + 1)^{1/2}$. With $I > 1/2$, a larger number of orientations (i.e., energy levels) are possible. The strength of the external magnetic field, $H_{applied}$, and the values of μ determine the energies and the energy difference between orientations:

$$\Delta E = h\nu = 2 \ \mu H_{applied} \qquad (2)$$

(where ν is the frequency of electromagnetic radiation absorbed or emitted). For the two possible orientations: one is of lower energy where the axis of the nuclear magnet is oriented exactly *parallel* or "with" the external magnetic field; the other is of higher energy where the axis of the nuclear magnet is oriented exactly *antiparallel* or "against" the external magnetic field (Fig. II.1).

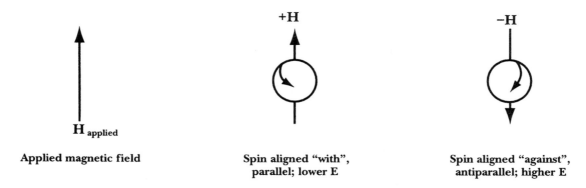

Figure II.I The two allowed spin states for nuclei of I = 1/2.

Now unless the axis of the nuclear magnet is oriented exactly parallel or antiparallel with the applied magnetic field, there will be a certain force by the external magnetic field to orient it. But because the nucleus is spinning, the effect is that its rotational axis draws out a circle perpendicular to the applied field. The nuclear motion is called *precession* (Fig. II.2); a gyroscope is an example.

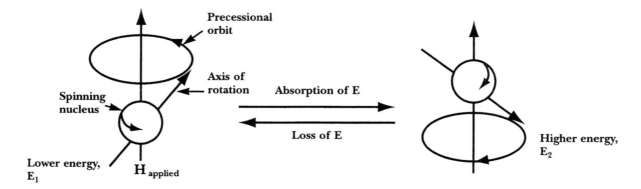

Figure II.2 Precession of the hydrogen nucleus.

It is possible to induce transitions between these two energy states and that is when the electromagnetic radiation of appropriate frequency is absorbed or emitted. The frequency of the electromagnetic radiation necessary for such a transition can be obtained rearranging equation 2:

$$\nu = \frac{2\,\mu H_{applied}}{h} \tag{3}$$

When the precessional frequency of the spinning nucleus is *exactly equal* to the frequency of electromagnetic radiation necessary for a transition to occur, that is, when the two frequencies are in *resonance*, then a transition from one nuclear spin state to another occurs. When in resonance at the applied magnetic field, the nuclei undergo a "spin flip", hence the name nuclear magnetic resonance (NMR) spectroscopy is used.

For a magnetic field of 14,100 gauss, the resonance frequency is in the radio-frequency (RF) range of the electromagnetic spectrum, 60 MHz. Since it is easier to vary the magnetic field than the RF, in most of the commercial equipment on the market, a set frequency is used (for example, 60 MHz or 200 MHz) and the magnetic field is varied by means of a current through a pair of sweep coils. The secondary field produced by the sweep coils is used to increase the field produced by the permanent magnets (Fig. II.3).

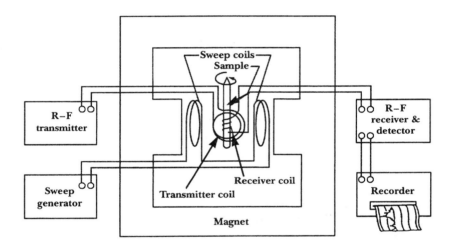

Figure II.3 Schematic diagram of an NMR spectrometer.

In an NMR experiment a sample is placed between the magnets of the instrument and the magnetic field is adjusted. When resonance is achieved the nucleus in the lower energy state E_1 absorbs electromagnetic energy and undergoes a "spin flip" into the higher energy state E_2. The absorption of energy can be detected electronically and recorded as a peak on a chart.

Chemical Shift

If a compound is put in a magnetic field that contains hydrogen atoms attached to different groups, the proton magnetic spectrum records the presence of various protons. For example, an NMR spectrum of benzyl acetate is shown in Fig. II.4. The peak with the highest intensity and at the lowest external field corresponds to the protons on the benzene ring; the least intense peak corresponds to the $-CH_2$ protons; and the peak at the highest external field corresponds to the $-CH_3$ protons. The intensity of the peak measures the number of protons experiencing the same magnetic field (see discussion of integration below).

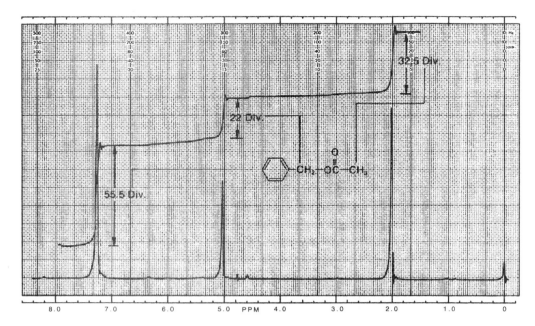

Figure II.4 Nuclear magnetic resonance spectrum of benzyl acetate; the absorption peak at the far right is caused by the added reference substance TMS.
(courtesy of Pavia, et al, Saunders College Publishing)

The reason that the benzene, CH_2 and CH_3 protons resonate at different resonance magnetic fields at a set RF is that each of these protons experiences a somewhat different magnetic field because of *electronic shielding*. Bare protons all experience the external magnetic field and, therefore, would resonate at the same frequency. However, the electronic environment of a nucleus in a magnetic field produces an induced magnetic field that opposes the applied field. Hence, the resultant field experienced by the proton is the *net* effect of the applied and the induced field brought about by the electrons shielding effect.

$$H_{at\ nucleus} = H_{applied} - H_{shielding} \qquad (4)$$

How much is the shielding? Shielding depends on the electrons in the bond between hydrogen and the atom to which it is attached. It is influenced by neighboring groups and atoms in the molecule, paramagnetism and other interatomic effects (see references at the end of this section for more details).

The resultant effect is that the protons in different electronic environments display spectral peaks at different fields. Thus, the NMR spectrum of benzyl acetate in Fig. II.4 shows three peaks which correspond to three different types of hydrogen atoms. Typical, also, is the display on the chart with the applied field increasing from left to right. Notice a small absorption peak at the far right. This corresponds to the reference compound, tetramethylsilane, $(CH_3)_4Si$, (TMS) which was added to the solution. This compound is chosen because it absorbs at a higher field strength than the hydrogen nuclei of nearly all other hydrogen containing compounds and thus, is used as a standard reference for NMR experiments.

The NMR spectrum is reported in terms of *chemical shift* from the reference compound. The chemical shift is a field dependent measure since only frequency differences are expressed. The chemical shift in delta, δ, expresses the amount by which a proton resonance is shifted from TMS.

$$\delta = \frac{H_r - H_i}{H_r} \times 10^6 \qquad (5)$$

It is calculated by equation 5, where H_i is the resonance field of the peak of interest, and H_r is the resonance field of the reference standard. The chemical shifts are multiplied by 10^6 and expressed in parts per million (ppm); 1 δ equals 1 ppm. Since the field strength and the resonance frequency are proportional, chemical shifts can also be given in terms of frequency (Hz) (equation 6); 1 δ equals 60 Hz.

$$\delta = \frac{\nu_r - \nu_i}{\nu_r} \times 10^6 \qquad (6)$$

Usually the chart papers supplied with commercial instruments have scales printed in both ppm and Hz terms. The resonance of TMS is at a higher field or "upfield" than that of most other types of protons and is assigned the value of 0 ppm; the chemical shifts reported with this reference value increase to the left or "downfield" from TMS. The methyl group protons in TMS experience the most shielding. As a proton becomes less shielded, hence more like an "acidic" or bare proton, its resonance frequency is shifted downfield. For the spectrum of benzyl acetate shown in Fig. II.4, the peaks for the indicated hydrogens have the following values: benzene ring hydrogens, 7.3 δ or 438 Hz; CH_2, 5.1 δ or 306 Hz; CH_3, 2.0 δ or 120 Hz.

A second standard scale for chemical shifts given in the literature is in terms of τ (tau) values. This scale also is independent of the oscillator frequency. In this scale TMS is assigned the value of 10 ppm. The values for chemical shift in this scale decrease as one moves downfield from TMS. The relationship of τ and δ are given in equation 7.

$$\tau = 10.00 - \delta \qquad (7)$$

For example, the benzene hydrogens for benzyl acetate in Fig. II.4 are at 2.7 γ (10.0 − 7.3 δ = 2.7 τ). The δ scale is more widely used.

The chemical shift for a given type of proton associated with a particular functional group is fairly constant. This means that if the NMR of an unknown organic compound is available, a chemist can gain valuable information about the types of hydrogens and the functional groups in the compound. Typical chemical shifts of protons are tabulated which correlate them with structural features. Extensive tables can be found in the references at the end of this section. Fig. II.5 contains a chart which compiles ranges over which most of the frequently found protons occur.

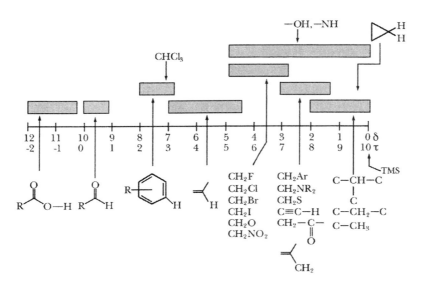

Figure II.5 Chart correlating chemical shifts for frequently found protons.
(courtesy Pavia, et al, Saunders College Publishing)

Integration

Other information is available from the NMR spectrum. The relative numbers of the different types of hydrogens can be determined by *integration*. This is possible since the area under the curve for each peak is proportional to the number of protons producing the peak. The electronics of the instrument produces a stepped or integral curve for each peak; the height of each curve is proportional to the number of protons. Thus, for the example in Fig. II.4, from the measurement of the vertical height of each stepped curve, the simple whole number ratio of the protons in the molecule is 5:2:3. If the molecular formula of the organic compound is known, these relative ratios can be expressed as the absolute numbers of hydrogens present in the molecule.

Spin-Spin Splitting

Another important piece of information that can be obtained from the NMR spectrum is the *spin-spin splitting*. Often an absorption peak is split into a collection of several smaller peaks. This results because there is magnetic interaction, or *coupling* (the details of which can be explored in the references following this section), between the proton responsible for the peak with a *magnetically non-equivalent* proton on the same carbon or on an adjacent carbon.

In the spectrum of ethyl bromide at low resolution, two peaks appear (Fig. II.6). The peak at lower intensity and the lower external field corresponds to the CH_2 protons, and the larger peak at the higher external field corresponds to the CH_3 protons.

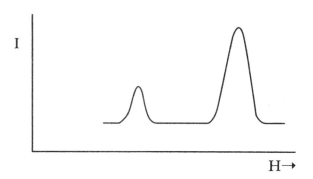

Figure II.6 Low resolution NMR of ethyl bromide.

There is little definition to the peaks. When one observes proton magnetic resonance in a high resolution apparatus, other features of the spectrum in addition to the chemical shifts become apparent. The peaks of the low resolution apparatus, usually broad, become more complex, and they may appear as singlets, doublets, triplets, and so forth. In the case of ethyl bromide, the low field peak appears as a group of four peaks, a symmetrical quartet, and the high field peak appears as three peaks, a symmetrical triplet (Fig. II.7). (The integration does not change, however, and is in the ratio 2:3.) The splitting of the peaks into multiplets is caused by electron-coupled spin-spin interactions of the protons in neighboring groups. The separation of the peaks within the multiplets is known as the spin-spin coupling constant, J, and is reported in Hz; *it is a value that is independent of the magnetic field, in contrast to the chemical shift.*

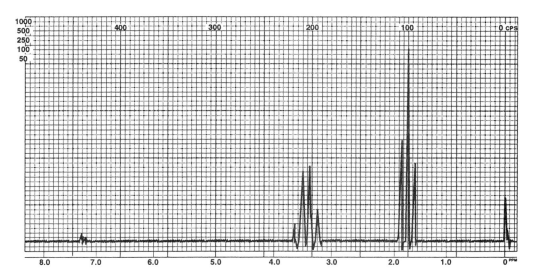

Figure II.7 High resolution NMR spectrum of ethyl bromide.
(courtesy of Varian Associates)

Spin-spin coupling and the resulting splitting obey a number of simple rules when: 1) the chemical shifts between the interacting groups are larger than the splitting due to the interaction; and 2) each proton in one group interacts equally with each proton in the other group.

1. Protons of an equivalent group (for example, the three protons of a methyl group) do not interact with each other in such a manner as to cause multiplets.

2. The multiplicity caused by a neighboring group is given by the equation **n + 1** where **n** is the number of neighboring protons.

The absorption by these two H atoms gives a quartet, a collection of four smaller peaks due to the three neighbors: n = 3, n + 1 = 3 + 1 = 4.

$$\left\{ \begin{array}{ccc} H & & H \\ | & & | \\ C & - & C - H \\ | & & | \\ H & & H \end{array} \right\}$$

The absorption by these three H atoms gives a triplet, a collection of three smaller peaks due to the two neighbors: n = 2, n + 1 = 2 + 1 = 3.

3. If more than one neighboring group interacts, the multiplicity of the resulting peak will be given by the product $(n_1 + 1)(n_2 + 1)$. For example, in the proton magnetic resonance spectrum of 1- chloropropane, $\overset{\gamma}{CH_3} - \overset{\beta}{CH_2} - \overset{\alpha}{CH_2} - Cl$, the β-methylene protons should appear in the high resolution spectrum as 12 peaks bunched together:
$n_\gamma = 3$, $n_\alpha = 2$, then $(3 + 1)(2 + 1) = 12$.

Chemical shifts, integrated intensities and spin-spin splitting also appear in the NMR spectra of other atoms, such as ^{13}C, ^{13}P, etc., albeit at different resonance frequencies.

Sample Preparation

An NMR spectrum is usually taken of a liquid sample. Depending on the sample and the solubility, a solution of 5% to 20% is prepared. Solvents usually used must be either deuterated (for example, use $CDCl_3$ rather than $CHCl_3$) or non-protonated (for example, CCl_4 or CS_2). The sample is placed in NMR tubes made of special thin-walled glass.

Solving a Problem Given the Spectrum

The NMR spectrum can provide significant information in the determination of the structure of an organic molecule. The three pieces of information most readily available from the spectrum follow.

1. **Chemical shift.** Determining the chemical shift of a single peak or a group of peaks enables the chemist to propose possible *functional groups* to which the protons producing the peaks are part.

2. **Spin-spin splitting.** The pattern and multiplicity of a peak provide information about the *number of nearest neighbors* to the proton giving the peak.

3. **Integration.** Integration of all the peaks gives the chemist information about the *relative abundance* of each type of proton present in the molecule. (Given the molecular formula, relative numbers can be converted into absolute numbers.)

Problems

Five problems follow. Each is the spectrum of an organic molecule whose structure can be decuced from the spectrum and the information given with the spectrum.

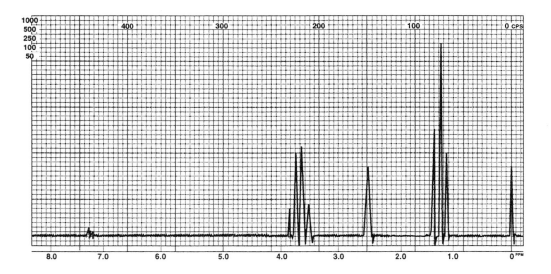

Problem II.1 C_2H_6O; integral ratio 2 : 1 : 3 (spectrum courtesy of Varian Associates).

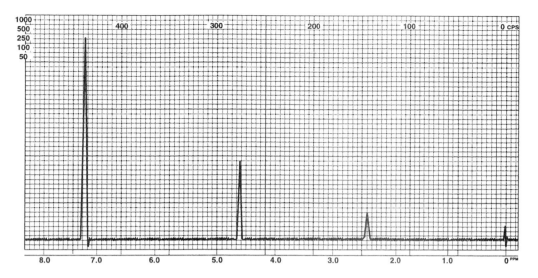

Problem II.2 C_7H_8O; integral ratio 5 : 2 : 1 (spectrum courtesy of Varian Associates).

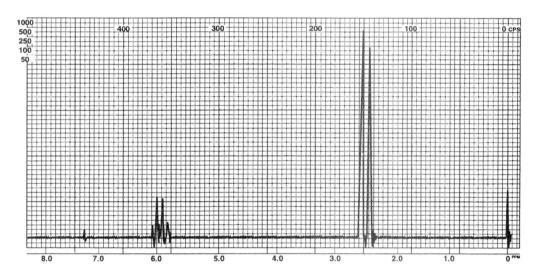

Problem II.3 $C_2H_4Br_2$; integral ratio 1 : 3 (spectrum courtesy of Varian Associates).

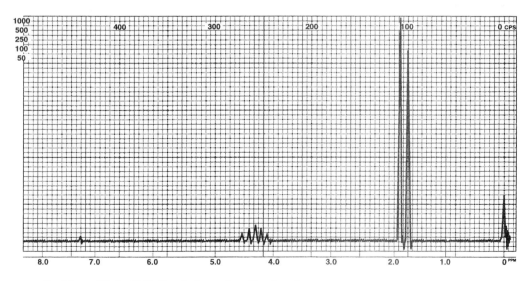

Problem II.4 C_3H_7Br; integral ratio 1 : 6 (spectrum courtesy of Varian Associates).

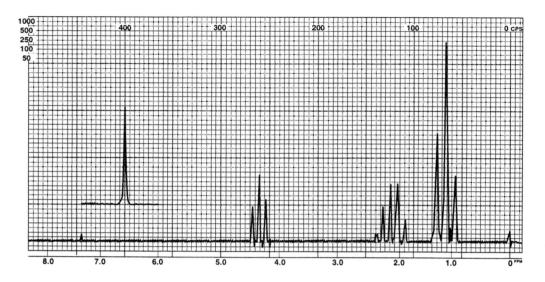

Problem II.5 $C_4H_7BrO_2$; integral ratio 1 : 1 : 2 : 3 (spectrum courtesy of Varian Associates).

References

Brown, W. H.,   Organic Chemistry, p. 475 - 520, Saunders College Publishing, Philadelphia, PA (1995).

Dyer, J. R.,   Applications of Absorption Spectroscopy of Organic Compounds, p. 58 - 132, Prentice-Hall, Inc., Englewood Cliffs, NJ (1965).

Robinson, J. W.,   Undergraduate Instrumental Analysis, p. 85 - 120, Marcel Dekker, Inc., New York, NY (1987).

Silverstein, R. M., Bassler, G. C., Morrill, T. C.,   Spectrometric Identification of Organic Compounds, 5th Ed., p. 165 - 287, John Wiley & Sons, Inc., New York, NY (1991).

Sorrell, T. H.,   Interpreting Spectra of Organic Molecules, p. 53 - 115, University Science Books, Mill Valley, CA (1988).

Appendix III

List of Apparatus and Equipment in Student's Locker

AMOUNT AND DESCRIPTION

- (1) Beaker, 50-mL
- (1) Beaker, 100-mL
- (1) Beaker, 250-mL
- (1) Beaker, 400-mL
- (1) Beaker, 600-mL
- (1) Clamp, test tube
- (1) Cylinder, graduated by 0.1 mL, 10-mL
- (1) Cylinder, graduated by 1 mL, 100-mL
- (1) Dropper, medicine with rubber bulb
- (1) Evaporating dish
- (1) File, triangular
- (1) Filtervac or neoprene adapter No. 2
- (1) Filter flask with side arm, 250-mL
- (1) Filter flask with side arm, 50-mL
- (1) Flask, Erlenmeyer, 125-mL
- (1) Flask, Erlenmeyer, 250-mL
- (1) Funnel, Büchner, 85 O.D.
- (1) Funnel, Hirsch
- (1) Funnel, short stem
- (1) Gauze, wire
- (1) Organic kit (glassware with 19/20 standard tapered joints: addition funnel, 125-mL; Claisen adapter; thermometer adapter; vacuum adapter; drying tube; water-cooled condenser; 3 round bottom flasks: 150-mL, 50-mL, and 10-mL)
- (1) Spatula, stainless steel
- (1) Sponge
- (1) Striker
- (6) Test tubes, 15 × 150 mm
- (6) Test tubes, 13 × 100 mm
- (1) Thermometer, 150°C
- (1) Tongs, crucible
- (1) Wash bottle, plastic
- (1) Watch glasses, 3-in. and 5-in.

List of Common Equipment and Materials in the Laboratory

Appendix IV

Each laboratory should be equipped with hoods and safety-related items such as fire extinguisher, fire blankets, safety shower, and eye wash fountain. The equipment and materials listed here for 25 students should be made available in each laboratory.

Acid tray
Aspirators (splashgun type) on sink faucet
Balances, triple beam (or centigram) or top-loading
Barometer
Boiling chips (sulfuric acid resistant)
Bunsen burners, with latex rubber tubing
Capillary tubes for melting point determinations
Clamps, extension
Clamps, thermometer
Clamps, utility
Containers for solid chemical waste disposal
Containers for liquid organic waste disposal
Detergent for washing glassware
Filter papers
Gas chromatograph (optional)
Glass rods, 4 and 6 mm OD
Glass tubing, 6 and 8 mm OD
Glycerol (glycerine) in dropper bottles
Infrared spectrophotometer (optional)
Heating mantles (100-mL and 50-mL) or thermowells
Hot plates/magnetic stirrers
Ice maker
Melting point apparatus
Paper towel dispensers
Pasteur pipets
Racks, test tube
Rings, support, iron, 76 mm OD
Ring stands
Rubber bulbs for Pasteur pipets
Rubber tubing, pressure
Rubber tubing, soft (0.25 in. OD)

Sample vials with screw caps, 2 drams capacity
Sea sand
Silicon grease for joints
Spectroline pipeters
Steam baths, copper
Variacs (voltage regulators)
Water, deionized or distilled
Weighing dishes, polystyrene, disposable, 73 × 73× 25 mm
Weighing paper

Special Equipment and Chemicals

Appendix V

In the instructions below every time a solution is to be made up in *"water"* you must use *distilled water*.

EXPERIMENT 1

STRUCTURE IN ORGANIC COMPOUNDS: USE OF MOLECULAR MODELS.I

Special Equipment

	(Color of spheres may vary depending on the set; substitute as necessary.)
(50)	Black spheres - 4 holes
(300)	Yellow spheres - 1 hole
(50)	Colored spheres (e.g. green) - 1 hole
(25)	Blue spheres - 2 holes
(400)	Sticks
(25)	Protractors
(75)	Springs (Optional)

EXPERIMENT 2

THE SEPARATION OF THE COMPONENTS OF A MIXTURE

Special Equipment

(2)	Top loading balance (weigh to 0.001 g)
[(15)	Centigram balance (weigh to 0.01 g) as an alternative]
(25)	Evaporating dish, porcelain, 6 cm dia.
(25)	Rubber policeman
(1 box)	Filter paper, 15 cm, fast flow
(25)	Mortar and pestle
(25)	Spatula

Chemicals

(30 g)	Unknown mixture: mix 3.0 g naphthalene (10%), 15 g sodium chloride (50%), 12 g sea sand (40%)
(1 jar)	Boiling stones (silicon carbide chips if available)

EXPERIMENT 3

RESOLUTION OF A MIXTURE BY DISTILLATION

Special Equipment

(25)	Distillation kits with 19/22 standard taper joints: 100-mL round bottom flasks (2); distilling head; thermometer adapter; 110°C thermometer; condenser; vacuum adapter; 25-mL round bottom flask; 10-mL round bottom flasks (4)
(1)	Gas chromatograph
(25)	Heating mantles, 100-mL capacity
(25)	Heating mantles, 25-mL capacity
(25)	Nickel wires
(1)	10-μL syringe
(25)	Variacs

Alternative heating sources for carrying out the distillation:

(25)	Thermowells, each of 25-mL and 100-mL capacity
	or
(25)	Hot plates to be used with sand baths; the sand bath can be made from a crystallizing dish (125 × 65 mm) to be filled half-way with sand.

Chemicals

(1 jar)	Silicone grease
(1 jar)	Boiling chips
(2 L)	5% sodium chloride solution: dissolve 100 g NaCl in enough water to make 2 L
(100 mL)	0.5 M silver nitrate: dissolve 8.5 g $AgNO_3$ in enough water to make 100 mL solution
(50 mL)	Concentrated nitric acid, HNO_3
(200 mL)	Cyclohexane
(200 mL)	Toluene

EXPERIMENT 4

PHYSICAL PROPERTIES OF CHEMICALS: MELTING POINT, SUBLIMATION AND BOILING POINT

Special Equipment

(1 roll)	Aluminum foil
(1 bottle)	Boiling chips
(1)	Commercial melting point apparatus (if available)
(25)	Glass tubing, 20 cm segments
(25)	Hot plates
(100)	Melting point capillary tubes
(50)	Rubber rings (cut 0.25-in. rubber tubing into narrow segments)
(25)	Thermometer clamps
(25)	Thiele tube melting point apparatus

Chemicals

(20 g)	Acetamide
(20 g)	Acetanilide
(20 g)	Benzophenone
(20 g)	Benzoic acid
(20 g)	Biphenyl
(20 g)	Lauric acid
(20 g)	Naphthalene, pure
(50 g)	Naphthalene, impure: mix 47.5 g (95%) naphthalene and 2.5 g (5%) charcoal powder.
(20 g)	Stearic acid

The following liquids should be placed in dropper bottles.

(200 mL)	Acetone
(200 mL)	Cyclohexane
(200 mL)	Ethyl acetate
(200 mL)	Hexane
(200 mL)	Isopropyl alcohol (2-propanol)
(200 mL)	Methyl alcohol (methanol)
(200 mL)	1-Propanol

EXPERIMENT 5

COLUMN AND PAPER CHROMATOGRAPHY: SEPARATION OF PLANT PIGMENTS

Special equipment

(50)	Melting point capillaries open at both ends
(25)	25-mL burets
(1 jar)	Glass wool
(25)	Filter papers (Whatman no.1), 20 × 10 cm
(3)	Heat lamp (optional)
(25)	Ruler with both English and metric scale
(1)	Stapler
(15)	Hot plates with or without water bath

Chemicals

(1 lb)	Tomato paste
(500 g)	Aluminum oxide (alumina)
(500 mL)	95% ethanol
(500 mL)	Petroleum ether, b.p. 35-60°C
(500 mL)	Eluting solvent: mix 450 mL petroleum ether with 10 mL toluene and 40 mL acetone.
(10 mL)	0.5% β-carotene solution: dissolve 50 mg in 10 mL petroleum ether. Wrap the vial in aluminum foil to protect from light and keep in refrigerator until used.
(150 mL)	Saturated bromine water
(500 mg)	Iodine crystals

EXPERIMENT 6

ISOLATION OF CAFFEINE FROM TEA LEAVES

Special Equipment

(25)	Cold finger condensers (115 mm long × 15 mm OD)
(1 box)	Filter paper; 7.0 cm, fast flow (Whatman no.1)
(25)	Hot plates
(50)	Latex tubing, 2 ft. lengths
(1 vial)	Melting point capillaries
(25)	No. 2 neoprene adaptors
(25)	Rubber stopper (no. 6, 1-hole) with glass tubing inserted (10 cm length × 7 mm OD)
(25)	125-mL separatory funnels
(25)	25-mL side-arm filter flasks
(25)	250-mL side-arm filter flasks
(25)	Small sample vials
(1)	Stapler
(50)	Vacuum tubing, 2 ft. lengths
(1 box)	Weighing paper

Chemicals

(1 bottle)	Boiling chips
(500 mL)	Methylene chloride, CH_2Cl_2
(25 g)	Sodium sulfate, anhydrous, Na_2SO_4
(50 g)	Sodium carbonate, anhydrous, Na_2CO_3
(50)	Tea bags

EXPERIMENT 7

IDENTIFICATION OF HYDROCARBONS

Special Equipment

(2 vials)	Litmus paper, blue
(250)	100 × 13 mm test tubes

Chemicals

(25 g)	Iron filings or powder

The following solutions should be placed in dropper bottles.

(100 mL)	Concentrated H_2SO_4
(100 mL)	Cyclohexene
(100 mL)	Hexane
(100 mL)	Ligroin (b.p. 90-110°C)
(100 mL)	Toluene

(100 mL) 1% Br_2 in cyclohexane (wear goggles and gloves; prepare under hood): mix 1.0 mL Br_2 with enough cyclohexane to make 100 mL. Prepare fresh solutions prior to use; keep in a dark brown dropper bottle; do not store.

(100 mL) 1% aqueous $KMnO_4$: dissolve 1.0 g potassium permanganate in 50 mL distilled water by gently heating for 1 hr.; cool and filter; dilute to 100 mL. Store in a dark brown dropper bottle.

(100 mL) Unknown A = hexane

(100 mL) Unknown B = cyclohexene

(100 mL) Unknown C = toluene

EXPERIMENT 8

DEHYDRATION OF 2-METHYLCYCLOHEXANOL; AN ELIMINATION REACTION

Special Equipment

(1 jar) Boiling stones, porous (for use with acid solutions; carbon chips or silicon carbide)

(25) Heating mantles: large enough to fit 50-mL and 25-mL round bottom flasks. The thermowell can be filled with sea sand in order to fit smaller flasks.

(50) Latex tubing, 2 ft. lengths

(1 box) Pasteur pipets

(25) Ring clamps

(25) 125-mL separatory funnels

(25) Variacs

(1 box) Vials, screw cap, 2 dram capacity

(25) Distillation kits, 19/20 standard taper, consisting of: round bottom flasks [50-mL(1), 25-mL(2)]; distillation head; thermometer adapter; thermometer, 150°C; vacuum adapter; water-cooled condenser.

Optional Equipment

(1) 10- μL syringe

(1) Set of sodium chloride, NaCl, discs and holder

(1) Infrared spectrophotometer

(1) Gas chromatograph

Chemicals

(50 mL) Cyclohexane

(50 mL) 1,2-Dimethoxyethane

(400 mL) 2-Methylcyclohexanol

(200 mL) Bromine solution: a 1% solution can be prepared with any of the following solvents: cyclohexane, carbon tetrachloride, or methylene chloride; the test works with all three; use discretion in your choice. Mix 2.0 mL Br_2 with enough solvent to make 200 mL. Prepare in hood; wear goggles and gloves. Prepare fresh solutions prior to use; keep in dark brown dropper bottle; do not store.

(150 mL) 85% phosphoric acid, H_3PO_4

(200 mL) Potassium permanganate, aqueous, $KMnO_4$: a 1% solution can be prepared by dissolving 2.0 g potassium permanganate in 100 mL distilled water by gently heating for 1 hr.; cool, filter and dilute to 200 mL. Store in a dark brown dropper bottle.

(1 L) 10% sodium carbonate, Na_2CO_3: dissolve 100 g sodium carbonate, anhydrous, in enough distilled water to make 1 L.

(50 g) Sodium sulfate, anhydrous, granular, Na_2SO_4

EXPERIMENT 9

IDENTIFICATION OF ALCOHOLS AND PHENOLS

Special Equipment

(125) Corks (for test tubes 100×13 mm)

(125) Corks (for test tubes 150×18 mm)

(25) Hot plate

(5 rolls) Indicator paper (pH 1 - 12)

Chemicals

The following solutions should be placed in dropper bottles.

(100 mL) Acetone (reagent grade)

(100 mL) 1-Butanol

(100 mL) 2-Butanol

(100 mL) 2-Methyl-2-propanol (t-butyl alcohol)

(200 mL) 20% aqueous phenol: dissolve 80 g of phenol in 20 mL distilled water; dilute to 400 mL

(100 mL) Lucas reagent (prepare under hood; wear goggles and gloves): cool 100 mL of concentrated HCl in an ice bath; with stirring, add 150 g anhydrous $ZnCl_2$ to the cold acid.

(150 mL) Chromic acid solution (prepare under hood; wear goggles and gloves): dissolve 30 g chromic oxide, CrO_3, in 30 mL concentrated H_2SO_4. Carefully add this solution to 90 mL water.

(100 mL) 2.5% ferric chloride solution: dissolve 2.5 g anhydrous $FeCl_3$ in 50 mL water; dilute to 100 mL.

(100 mL) Iodine in KI solution: mix 20 g of KI and 10 g of I_2 in 100 mL water.

(250 mL) 6 M sodium hydroxide, 6 M NaOH: dissolve 60 g NaOH in 100 mL water. Dilute to 250 mL with water.

(100 mL) Unknown A = 1-butanol

(100 mL) Unknown B = 2-butanol

(100 mL) Unknown C = 2-methyl-2-propanol (t-butyl alcohol)

(100 mL) Unknown D = 20% aqueous phenol

EXPERIMENT 10

STEREOCHEMISTRY: USE OF MOLECULAR MODELS. II

Special Equipment

Commercial molecular model kits vary in style, size, material composition and the color of the components. The set which works best in this exercise is the *Molecular Model Set for Organic Chemistry* available from Allyn and Bacon, Inc. (Newton, MA). Wood ball and stick models work as well. For 25 students, 25 of these sets should be provided. If you wish to make up your own kit, you would need the following for 25 students:

(25)	Cyclohexane model kits: each consisting of the following components:
	8 carbons - black, 4-hole
	18 hydrogens - white, 1-hole
	2 substituents - red, 1-hole
	24 connectors - bonds
(25)	Chiral model kits: each consisting of the following components:
	8 carbons - black, 4-hole
	32 substituents - 8 red, 1-hole; 8 white, 1-hole; 8 blue, 1-hole; 8 green, 1-hole
	28 connectors - bonds
(5)	Small hand mirrors

EXPERIMENT 11

REDUCTION OF -DIKETONES BY YEAST

Special equipment

(1)	Gas-liquid partition chromatography apparatus
(1)	Infrared spectrophotometer
(15)	Silica gel TLC plates 10 × 4 cm
(15)	Rulers (1 mm divisions)
(1 box)	Polyethylene gloves
(50)	Capillary tubes open on both ends
(2)	Heat lamps or hair dryers
(1)	Drying oven, 120°C
(1)	UV lamp, 254 nm
(15)	Büchner funnels (85-mm OD)
(15)	Filtervacs or neoprene adaptors
(15)	1-L filter flasks
(15)	250-mL separatory funnel
(15)	25-mL burets
(15)	Evaporating dishes

Chemicals

(10 g)	1-phenyl-1,2-propanedione
(600 g)	Freeze-dried baker's yeast

(10 mL)   0.1% 1-phenyl-1,2-propanedione solution: dissolve 10 mg 1-phenyl-1,2-propanedione in 10 mL cyclohexane.

(150 mL)   Cyclohexane/diethyl ether mixture: add 75 mL of diethyl ether to 75 mL of cyclohexane.

(200 mL)   Vanillin spray: dissolve 6 g of vanillin in 200 mL absolute ethanol. With slow stirring add 1 mL concentrated sulfuric acid.

(100 g)   Celite filter aid

(1 L)   Diethyl ether

(2.5 kg)   Sodium chloride

(500 g)   Magnesium sulfate

(500 mL)   Dichloromethane

(50 g)   Silica gel 60 (Merck) 100-230 mesh

(100 mL)   Cyclohexane

(50 mL)   Nujol (mineral oil)

EXPERIMENT 12

PREPARATION OF 1-BROMOBUTANE (n-BUTYL BROMIDE)

Special Equipment

(1 jar)   Boiling chips, porous (for use with sulfuric acid solutions; carbon chips or silicon-carbide)

(50)   Funnels, glass

(25)   Heating mantles: large enough to fit 50-mL and 25-mL round bottom flasks. The thermowell can be filled with sea sand in order to fit smaller flasks.

(50)   Latex tubing, 2-ft. lengths

(1 box)   Pasteur pipets

(25)   Ring clamps

(1 jar)   Sea sand

(25)   125-mL separatory funnels

(1 jar)   Silicone grease

(25)   Variacs

(1 box)   Vials, screw cap, 2 dram capacity

(25)   Reflux/distillation kits, 19/22 standard taper, consisting of: round bottom flasks [50-mL(1), 25-mL(2), 10-mL(1)]; distillation head; thermometer adapter; thermometer, 150°C; vacuum adapter; water-cooled condenser.

Optional equipment

(1)   10-μL syringe

(1)   Set of sodium chloride, NaCl, discs and holder

(1)   Infrared spectrophotometer

(1)   Gas chromatograph

Chemicals

(300 mL)   1-Butanol (n-butyl alcohol), $CH_3CH_2CH_2CH_2OH$

(25 g)   Sodium bisulfite, $NaHSO_3$

(350 g)   Sodium bromide, NaBr

(300 mL) 10% sodium carbonate, Na_2CO_3: dissolve 30 g sodium carbonate, anhydrous, in enough distilled water to make 300 mL.

(50 g) Sodium sulfate, anhydrous, granular, Na_2SO_4

(300 mL) Concentrated sulfuric acid, H_2SO_4

(300 mL) 9 M sulfuric acid, 9 M H_2SO_4: with stirring, carefully pour 150 mL concentrated sulfuric acid, H_2SO_4, into 100 mL ice water. Dilute to 300 mL volume. Wear rubber gloves, a rubber apron and a face shield when preparing.

EXPERIMENT 13

A FRIEDEL-CRAFTS ALKYLATION

Special Equipment

(25) Calcium chloride drying tubes, 19/20 standard taper

(25) Claisen adapters, 19/20 standard taper

(25) Crystallizing dishs, Pyrex, 125 × 65 mm

(25) Magnetic spin-bars, 0.5 in.

(25) Magnetic stirrer/hot plates

(1 box) Pasteur pipets

(25) Ring clamps

(25) 25-mL round bottom flasks, 19/20 standard taper

(25) Rubber septa

(25) 125-mL separatory funnels

(25) Syringes, 1-mL

(1) Syringe, 10-μL

(25) Thermometers, 110°C

(1 box) Vials, 2 dram capacity

Chemicals

(10 g) Aluminum chloride, anhydrous, $AlCl_3$

(100 g) Calcium chloride, anhydrous, $CaCl_2$

(2.5 Kg) Sea sand

(100 g) Sodium sulfate, granular, anhydrous, Na_2SO_4

The following should be put in dropper bottles.

(200 mL) 5% aqueous sodium bicarbonate, $NaHCO_3$: dissolve 10 g of sodium bicarbonate in 100 mL of distilled water; add enough distilled water to bring to 200 mL.

(100 mL) <u>n</u>-Propyl chloride (1-propanol)

(200 mL) <u>p</u>-Xylene

EXPERIMENT 14

PROPERTIES OF AMINES AND AMIDES

Special Equipment

(2 rolls)   pH paper (range 0 to 12)
(25)   Hot plates

Chemicals

(20 g)   Acetamide

The following solutions should be placed in dropper bottles.

(25 mL)   Triethylamine
(25 mL)   Aniline
(25 mL)   N,N-Dimethylaniline
(100 mL)   Diethyl ether (ether)
(100 mL)   6 M ammonia solution, 6 M NH_3: 40 mL concentrated NH_4OH diluted to 100 mL with water; prepare in the hood.
(100 mL)   6 M hydrochloric acid, 6 M HCl: 50 mL concentrated HCl diluted to 100 mL with water. Wear rubber gloves, a rubber apron and a face shield when preparing; do in the hood.
(50 mL)   Concentrated hydrochloric acid, HCl
(250 mL)   6 M sulfuric acid, 6 M H_2SO_4: pour 82.5 mL concentrated H_2SO_4 into 125 mL cold water. Stir slowly. Dilute to 250 mL with water. Wear rubber gloves, a rubber apron and a face shield when preparing; do in the hood.
(250 mL)   6 M sodium hydroxide, 6 M NaOH: dissolve 60 g NaOH in 100 mL water. Dilute to 250 mL with water. Do in the hood.

EXPERIMENT 15

IDENTIFICATION OF ALDEHYDES AND KETONES

Special Equipment

(250)   Corks (to fit 100 x 13 mm test tube)
(125)   Corks (to fit 150 x 18 mm test tube)
(1 box)   Filter paper (students will need to cut to size)
(25)   Hirsch funnels
(25)   Hot plates
(25)   Neoprene adapters (no. 2)
(25)   Rubber stopper assemblies: a no. 6 one-hole stopper fitted with glass tubing (15 cm in length $\times$ 7 mm OD)
(25)   50-mL side-arm filter flasks
(25)   250-mL side-arm filter flasks
(50)   Vacuum tubing, heavy-walled (2 ft. lengths)

Chemicals

(50 g)   Hydroxylamine hydrochloride
(100 g)   Sodium acetate

The following solutions should be placed in dropper bottles.

(100 mL)   Acetone (reagent grade)
(100 mL)   Benzaldehyde (freshly distilled)
(100 mL)   Bis(2-ethoxymethyl) ether
(100 mL)   Cyclohexanone
(500 mL)   Ethanol (absolute)
(500 mL)   Ethanol (95%)
(100 mL)   Isovaleraldehyde
(500 mL)   Methanol
(100 mL)   Pyridine
(150 mL)   Chromic acid reagent: dissolve 30 g chromic oxide (CrO_3) in 30 mL concentrated H_2SO_4. Add carefully to 90 mL water. Wear rubber gloves, a rubber apron and a face shield during the preparation. Do in the hood.
           Tollens' reagent
           (100 mL)   Solution A: dissolve 9.0 g silver nitrate in 90 mL of water; dilute to 100 mL.
           (100 mL)   Solution B: 10 g NaOH dissolved in enough water to make 100 mL.

(100 mL)   10% ammonia water: 33.3 mL of 30% reagent grade ammonium hydroxide diluted to 100 mL.
(100 mL)   6 M sodium hydroxide, 6 M NaOH: dissolve 24 g NaOH in enough water to make 100 mL.
(500 mL)   Iodine-KI solution: mix 100 g of KI and a 50 g of iodine in enough distilled water to make 500 mL.
(100 mL)   2,4-Dinitrophenylhydrazine reagent: dissolve 3.0 g of 2,4-dinitrophenylhydrazine in 15 mL concentrated H_2SO_4. In a beaker, mix together 10 mL water and 75 mL 95% ethanol. With vigorous stirring slowly add the 2,4-dinitrophenylhydrazine solution to the aqueous ethanol mixture. After thorough mixing, filter by gravity through a fluted filter paper. Wear rubber gloves, a rubber apron and a face shield during the preparation. Do in the hood.
(100 mL)   Semicarbazide reagent: dissolve 22.2 g of semicarbazide hydrochloride in 100 mL of distilled water.
(100 mL)   Unknown A = isovaleraldehyde
(100 mL)   Unknown B = benzaldehyde
(100 mL)   Unknown C = cyclohexanone
(100 mL)   Unknown D = acetone.

Additional compounds for use as unknowns:

Aldehydes
(100 mL)   2-Butanal (crotonaldehyde)
(100 mL)   Octanal (caprylaldehyde)
(100 mL)   Pentanal (valeraldehyde)

Ketones
(100 mL)   Acetophenone
(100 mL)   Cyclopentanone
(100 mL)   2-Pentanone
(100 mL)   3-Pentanone

EXPERIMENT 16

CARBOHYDRATES

Special Equipment

(50)   Medicine droppers
(125)   Microtest tubes or 25 white spot plates
(2 rolls)   Litmus paper3, red

Chemicals

(20 g)   Boiling chips
(400 mL)   Fehling's reagent (solution A and B, from Fisher Scientific Co.)
(200 mL)   3 M NaOH: dissolve 24 g NaOH in water and bring it to 200 mL volume
(200 mL)   2% starch solution: place 4 g soluble starch in a beaker. With vigorous stirring, add
            10 mL water to form a thin paste. Boil 190 mL water in another beaker. Add the starch
            paste to the boiling water and stir until the solution becomes clear. (store in a dropper bottle).
(200 mL)   2% sucrose: dissolve 4 g sucrose in 200 mL water
(200 mL)   3 M sulfuric acid: add 33 mL concentrated H_2SO_4 to 150 mL ice cold water; pour the
            sulfuric acid slowly along the walls of the beaker, this way it will settle on the bottom
            without much mixing; stir slowly in order not to generate too much heat; when fully
            mixed bring the volume to 200 mL.

(100 mL)   2% fructose: dissolve 2 g fructose in 100 mL water (store in a dropper bottle).
(100 mL)   2% glucose: dissolve 2 g glucose in 100 mL water (store in a dropper bottle).
(100 mL)   2% lactose: dissolve 2 g lactose in 100 mL water (store in a dropper bottle).
(100 mL)   0.01 M iodine in KI: dissolve 1.2 g KI in 80 mL water. Add 0.25 g I2. Stir until the iodine
            dissolves. Dilute the solution to 100 mL volume. (store in a dark dropper bottle).

EXPERIMENT 17

PROPERTIES OF CARBOXYLIC ACIDS AND ESTERS

Special Equipment

(5 rolls)	pH paper (range 1 - 12)
(100)	Disposable Pasteur pipets
(5 vials)	Litmus paper, blue
(25)	Hot plates

Chemicals

(10 g)	Salicylic acid
(10 g)	Benzoic acid

The following solutions are placed in dropper bottles.

(75 mL)	Acetic acid
(50 mL)	Formic acid
(25 mL)	Benzyl alcohol
(50 mL)	Ethanol (ethyl alcohol)
(25 mL)	2-Methyl-1-propanol (isobutyl alcohol)
(25 mL)	3-Methyl-1-butanol (isopentyl alcohol)
(50 mL)	Methanol (methyl alcohol)
(25 mL)	Methyl salicylate
(250 mL)	6 M hydrochloric acid, 6 M HCl: take 150 mL of concentrated HCl and add to 50 mL of cold water; dilute with enough water to 250 mL. Wear rubber gloves, a rubber apron and a face shield during the preparation. Do in the hood.
(100 mL)	3 M hydrochloric acid, 3 M HCL: take 50 mL 6 M HCl and bring to 100 mL.
(300 mL)	6 M sodium hydroxide, 6 M NaOH: dissolve 72 g NaOH in enough water to bring to 300 mL.
(150 mL)	2 M sodium hydroxide, 2 M NaOH: take 50 mL 6 M NaOH and bring to 150 mL.
(25 mL)	Concentrated sulfuric acid, H_2SO_4

EXPERIMENT 18

PREPARATION OF ISOPENTYL ACETATE (BANANA OIL)

Special Equipment

(1 jar)	Boiling chips, porous (for use with sulfuric acid solutions; carbon chips or silicon-carbide)
(25)	Heating mantles: large enough to fit 50-mL and 25-mL round bottom flasks. The thermowell can be filled with sea sand in order to fit smaller flasks.
(1 box)	Pasteur pipets
(25)	Ring clamps
(25)	125-mL separatory funnels
(1)	10-μL syringe
(25)	Variacs

(1 box)   Vials, screw cap, 2 dram capacity

(25)   Reflux/distillation kits, 19/22 standard taper, consisting of: round bottom flasks [50-mL(1), 25-mL(2)]; calcium chloride drying tube; distillation head; thermometer adapter; thermometer, 150°C; vacuum adapter; water-cooled condenser.

Chemicals

(100 g)   Calcium chloride, anhydrous, $CaCl_2$

(100 mL)   Concentrated sulfuric acid, H_2SO_4

(200 mL)   Glacial acetic acid

(300 mL)   Isopentyl alcohol

(2 L)   5% sodium bicarbonate, aqueous, $NaHCO_3$: dissolve 50 g of anhydrous sodium bicarbonate in 500 mL of distilled water. Dilute to 1 L volume.

(100 g)   Sodium sulfate, anhydrous, granular, Na_2SO_4

EXPERIMENT 19

POLYMERIZATION REACTIONS

Special Equipment

(25)   Hot plates

(25)   Cylindrical paper rolls or sticks

(25)   Bent wire approximately 10-cm long

(25)   10-mL pipets or syringes

(25)   Spectroline pipet fillers

(25)   Beaker tongs

Chemicals

(750 g)   Styrene

(250 mL)   Xylene

(75 mL)   t-Butyl peroxide benzoate (also called t-Butyl benzoyl peroxide). Store at 4°C.

(75 mL)   20% NaOH: dissolve 15 g NaOH in 75 mL water.

(300 mL)   5% adipoyl chloride: dissolve 15 g adipoyl chloride in 300 mL cyclohexane.

(300 mL)   5% hexamethylene diamine: dissolve 15 g hexamethylene diamine in 300 mL water.

(200 mL)   80% formic acid: add 40 mL water to 160 mL formic acid.

EXPERIMENT 20

USE OF THE GRIGNARD REAGENT TO PREPARE BENZOIC ACID

Special Equipment

(25)	Büchner funnels (85 mm OD)
(25)	Heating mantles, 100-mL
(25)	Hot plates
(5 rolls)	pH paper, 2 - 5 range
(25)	250-mL side arm filter flasks
(25)	125-mL separatory funnel
(1 jar)	Silicone grease
(25)	Steam baths, copper
(25)	Variacs
(1 box)	Vials, screw cap, 2 dram capacity
(25)	Grignard kits, 19/20 standard taper, consisting of

> (1) 125-mL addition funnel
> (1) Claisen adapter
> (2) drying tubes
> (1) 100-mL round bottom flask
> (1) water-cooled condenser

Chemicals

(200 mL)	Bromobenzene
(500 g)	Calcium chloride, $CaCl_2$, anhydrous, 4 - 20 mesh
(3 cans)	Diethyl ether, anhydrous, (3×1 L cans)
(250 g)	Dry Ice
(500 mL)	Concentrated hydrochloric acid, HCl
(2 L)	9 M hydrochloric acid, 9 M HCl: with stirring, carefully pour 1500 mL concentrated hydrochloric acid, HCl, into 500 mL ice water. Wear rubber gloves, a rubber apron and a face shield when preparing.
(10 g)	Iodine, crystals
(35 g)	Magnesium, Mg, ribbon; 25 pieces approx. 161 cm in length
(3 L)	5% aqueous sodium hydroxide, 5% NaOH: dissolve 150 g sodium hydroxide, NaOH, pellets in one liter distilled water; dilute with enough distilled water to 3 liters.
(200 g)	Sodium sulfate, Na_2SO_4, anhydrous, granular

EXPERIMENT 21

PREPARATION OF ACETYLSALICYLIC ACID (ASPIRIN)

Special Equipment

(25)	Büchner funnels (85-mm OD)
(25)	Filtervac or neoprene adapters
(1 box)	Filter paper (7.0 cm, Whatman no. 2)
(25)	250-mL filter flasks
(25)	Hot Plates

Chemicals

(1 bottle) Boiling chips
(25)   Commercial aspirin tablets
(100 mL)   Concentrated sulfuric acid, H_2SO_4 (in a dropper bottle)
(100 mL)   1% ferric chloride: dissolve 1 g $FeCl_3 \cdot 6 H_2O$ in enough distilled water to make 100 mL (in a dropper bottle).
(100 mL)   Acetic anhydride, freshly opened bottle
(300 mL)   95% ethanol
(100 g)   Salicylic acid

EXPERIMENT 22

SYNTHESIS OF SULFANILAMIDE; A MULTISTEP SYNTHESIS

Special equipment

(25)   Claisen adapters
(25)   100-mL round bottom flasks
(25)   125-mL separatory funnels
(25)   250-mL beakers
(25)   Büchner funnels (85-mm OD)
(25)   Filtervac
(1 box)   Filter paper (7.0 cm, Whatman no. 2)
(25)   Hot plates with steam baths
(25)   Reflux condensers, water-cooled
(2)   Melting point apparatus
(25)   250-mL filter flasks
(2 vials) Melting point capillaries
(2 rolls) Litmus paper red
(2 rolls) Congo red test paper

Chemicals

(50 g)   Acetanilide
(300 mL)   Chlorosulfonic acid
(500 mL)   Aqueous ammonia 28%: this is labeled as concentrated ammnium hydroxide.
(500 mL)   6 M sulfuric acid: add 168 mL concentrated H_2SO_4 (95%) *slowly with constant stirring* to 250 mL ice-cold water and bring it to 500 mL volume. Wear rubber gloves, a rubber apron and a face mask during the preparation.
(500 mL)   5 M hydrochloric acid: add 215 mL concentrated HCl (36%) *slowly with constant stirring* to 250 mL ice-cold water. Mix and bring it to 500 mL volume. Wear rubber gloves, a rubber apron and a face mask during the preparation.
(500 g)   Sodium carbonate, Na_2CO_3.

EXPERIMENT 23

PREPARATION OF METHYL ORANGE

Special Equipment

(25)	Büchner funnels, 50 mm OD (or Hirsch funnels)
(25)	Filter flasks, 250-mL
(1 box)	Filter paper, Whatman no.1, 4.25 cm
(25)	Hot plates
(1 box)	Pasteur pipets
(2 rolls)	pH paper (range 1 - 12)
(50)	Pipets, 2-mL, graduated in 0.1 mL
(25)	Spectroline pipet filler
(25)	Test tubes (13 × 100)
(25)	Vacuum tubing, 2 ft. lengths
(1 box)	Vials, screw cap, 2 dram capacity

Chemicals

(50 mL)	N,N-Dimethylaniline
(50 mL)	Glacial acetic acid
(75 g)	Sulfanilic acid monohydrate
(100 mL)	Concentrated hydrochloric acid, HCl
(100 mL)	1 M Hydrochloric acid, 1 M HCl: mix 3.3 mL concentrated HCl with 50 mL cold water. Dilute by adding enough water to 100 mL. Wear rubber gloves, a rubber apron and a face shield during the preparation. Do in the hood.
(1 L)	2.5% aqueous sodium carbonate, Na_2CO_3: dissolve 25 g of sodium carbonate in 100 mL of water. Dilute by adding enough water to 1 L.
(1 L)	Saturated aqueous sodium chloride, NaCl: add 290 g sodium chloride, NaCl, to warm water (60°C). Stir until dissolved. Cool to room temperature.
(1 L)	10% sodium hydroxide, NaOH: dissolve 100 g of sodium hydroxide in 500 mL of water. Dilute by adding enough water to 1 L.
(100 mL)	1 M sodium hydroxide, 1 M NaOH: dissolve 4.0 g sodium hydroxide, NaOH, in 50 mL of water. Dilute by adding enough water to 100 mL.
(30 g)	Sodium nitrite, $NaNO_2$

EXPERIMENT 24

ISOCITRATE DEHYDROGENASE

Special Equipment

(15)	Büchner funnel (85 mm OD) in no. 7 one-hole rubber stopper
(15)	200-mL filter flasks
(1 box)	Filter papers, Whatman no.1

Chemicals

(30 g)	Baker's yeast
(100 g)	Acid washed sand
(200 mL)	0.1 M $NaHCO_3$ buffer: weigh 1.68 g sodium bicarbonate add distilled water to bring the volume to 200 mL.
(40 mL)	Phosphate buffer at pH 7.0: mix 25 mL of 0.1 M KH_2PO_4 and 15 mL of 0.1 M NaOH solutions. To prepare 0.1 M NaOH weigh 0.2 g NaOH, dissolve it in 20 mL distilled water and transfer it to a 50-mL volumetric flask and bring it to volume. To prepare 0.1 M KH_2PO_4 weigh 0.68 g of potassium dihydrogen phosphate, add sufficient distilled water to dissolve it and bring it to 50 mL volume.
(20 mL)	0.1 M $MgCl_2$ solution: dissolve 0.19 g magnesium chloride in 20 mL distilled water.
(20 mL)	2.5 mM ADP solution: dissolve 23 mg ADP in 20 mL distilled water.
(20 mL)	2.0 mM NAD^+ solution: dissolve 24 mg NAD^+ in 20 mL distilled water.
(50 mL)	5 mM sodium isocitrate solution: dissolve 53.5 mg sodium isocitrate in 50 mL distilled water.
(50 g)	Celite filter aid

EXPERIMENT 25

PREPARATION OF A HAND CREAM

Special Equipment

(25)	Bunsen burners

Chemicals

(100 mL)	Triethanolamine
(40 mL)	Propylene glycol (1,2-propanediol)
(500 g)	Stearic acid
(40 g)	Methyl stearate
(400 g)	Lanolin
(400 g)	Mineral oil

EXPERIMENT 26

ISOLATION OF LIPIDS FROM EGG YOLK

Special Equipment

(12)	Hard boiled eggs
(13)	Steam baths
(13)	Hot plates
(2)	Waste jars

Chemicals

(4 L)	Acetone
(1.5 L)	Diethyl ether (ether)

EXPERIMENT 27

ANALYSIS OF LIPIDS

Special Equipment

(75)	Capillary melting point tubes
(5)	Melting point apparatus
(25)	Hot plates

Chemicals

(250 mL)   Molybdate solution: dissolve 0.8 g $(NH_4)_6Mo_7O_{24}\cdot4H_2O$ in 30 mL water. Put in an ice bath. Slowly pour 20 mL concentrated sulfuric acid into the solution and stir it slowly. After cooling to room temperature bring the volume to 250 mL. Wear rubber gloves, a rubber apron and a face shield when preparing.

(50 mL)   0.1 M ascorbic acid solution: dissolve 0.88 g ascorbic acid (vitamin C) in 25 mL water and bring to 50 mL volume. This must be prepared fresh every week and stored at 4°C.

(250 mL)   6 M NaOH: dissolve 60 g NaOH in 100 mL water and bring the volume to 250 mL.

(250 mL)   6 M HNO_3: pipet 63 mL concentrated HNO_3 into a 250-mL volumetric flask containing 100 mL water; bring to volume with water. Wear safety equipment as described above.

(200 mL)   Chloroform

(75 mL)   Acetic anhydride

(50 mL)   Concentrated sulfuric acid, H_2SO_4

(75 g)   Potassium hydrogen sulfate, $KHSO_4$

EXPERIMENT 28

EXTRACTION AND IDENTIFICATION OF FATTY ACIDS FROM CORN OIL

Special equipment

(12)	Water bath
(2)	Heat lamps or hair dryers
(25)	15 × 6.5 cm silica gel TLC plates
(25)	Rulers, metric scale
(25)	Polyethylene, surgical gloves
(150)	Capillary tubes, open on both ends
(1)	Drying oven, 110°C

Chemicals

(50 g)	Corn oil
(5 mL)	Methyl palmitate solution: dissolve 25 mgmethyl palmitate in 5 mL petroleum ether.
(5 mL)	Methyl oleate solution: dissolve 25 mg methyl oleate in 5 mL petroleum ether.
(5 mL)	Methyl linoleate solution: dissolve 25 mg methyl linoleate in 5 mL petroleum ether.
(100 mL)	0.5 M KOH: dissolve 2.81 g KOH in 25 mL water and add 75 ml of 95 % ethanol.
(500 g)	Sodium sulfate, Na_2SO_4 , anhydrous, granular
(100 mL)	Concentrated hydrochloric acid, HCl
(1 L)	Petroleum ether (b.p. 30 - 60°C)
(300 mL)	Methanol:perchloric acid mixture: mix 285 mL methanol with 15 mL $HClO_4 \cdot 2H_2O$ (73 % perchloric acid).
(400 mL)	Hexane:diethyl ether mixture: mix 320 mL hexane with 80 mL diethyl ether.
(10 g)	Iodine crystals, I_2

EXPERIMENT 29

ANALYSIS OF VITAMIN A IN MARGARINE

Special equipment

(1)	UV spectrophotometer with suitable UV light source. Preferably it should be able to read down to 200 nm.
(1 pair)	Matched quartz cells, with 1 cm internal path length.
(2)	Long wavelength UV lamp. The lamp should provide radiation in the 300 nm range. (For example, #UVSL-55; LW 240 from Ultraviolet Products Inc.)
(12)	500-mL separatory funnel
(12)	25-mL burets
(12)	Hot plates with water baths
(12)	Beaker tongs

Chemicals

(0.5 lb)	Margarine
(400 mL)	50% KOH: weigh 200 mL KOH and add 200 mL water with constant stirring.
(1 L)	95% ethanol
(150 mL)	Absolute ethanol
(2 L)	Petroleum ether (b.p. 30 - 60°C)
(3 L)	Diethyl ether (practical)
(350 g)	Alkali aluminum oxide (alumina)

EXPERIMENT 30

TLC SEPARATION OF AMINO ACIDS

Special Equipment

(1)	Drying oven, 105 - 110°C
(2)	Heat lamps or hair dryers
(50)	15 × 6.5 cm silica gel TLC plates
(25)	Rulers, metric scale
(25)	Polyethylene, surgical gloves
(150)	Capillary tubes, open on both ends
(1 roll)	Aluminum foil

Chemicals

(25 mL)	0.12% aspartic acid solution: dissolve 30 mg aspartic acid in 25 mL distilled water.
(25 mL)	0.12% phenylalanine solution: dissolve 30 mg phenylalanine in 25 mL distilled water.
(25 mL)	0.12% leucine solution: dissolve 30 mg leucine in 25 mL distilled water.
(25 mL)	Aspartame solution: dissolve 150 mg Equal sweetener powder in 25 mL distilled water.
(50 mL)	3 M HCl solution: place 10 mL distilled water into a 50 mL volumetric flask. Add slowly 12.5 mL of concentrated HCl and bring it to volume with distilled water.
(1 L)	Solvent mixture: mix 600 mL n-butanol with 150 mL acetic acid and 240 mL distilled water.
(1 can)	Ninhydrin spray reagent (0.2% ninhydrin in ethanol or acetone); do not use any reagent older than 6 months old.
(1 can)	Diet Coca-Cola
(4 packets)	Equal, NutraSweet, sweeteners

EXPERIMENT 31

ISOLATION AND IDENTIFICATION OF CASEIN

Special Equipment

(25)	Hot plates
(25)	600-mL beakers
(25)	Büchner funnels (O.D. 85 mm) in No. 7 hole rubber stopper
(7 boxes)	Whatman No. 2 filter paper, 7 cm
(25)	Rubber bands
(25)	Cheese cloths (6 × 6 in.)

Chemicals

(25 g)	Boiling chips
(1 L)	95% ethanol
(1 L)	Diethyl ether:ethanol mixture (1:1)
(0.5 gal)	Regular milk
(500 mL)	Glacial acetic acid

The following solutions should be placed in dropper bottles:

(100 mL) Concentrated nitric acid, HNO_3, 69%
(100 mL) 2% albumin suspension: dissolve 2 g albumin in 100 mL water.
(100 mL) 2% gelatin: dissolve 2 g gelatin in 100 mL water.
(100 mL) 2% glycine: dissolve 2 g glycine in 100 mL water.
(100 mL) 5% copper(II) sulfate: dissolve 5 g $CuSO_4$ (or 7.85 g $CuSO_4 \cdot 5H_2O$) in 100 mL water.
(100 mL) 5% lead(II) nitrate: dissolve 5 g $Pb(NO_3)_2$ in 100 mL water.
(100 mL) 5% mercury(II) nitrate: dissolve 5 g $Hg(NO_3)_2$ in 100 mL water.
(100 mL) Ninhydrin reagent: dissolve 3 g ninhydrin in 100 mL acetone.
(100 mL) 10% sodium hydroxide: dissolve 10 g NaOH in 100 mL water.
(100 mL) 1% tyrosine: dissolve 1 g tyrosine in 100 mL water.
(100 mL) 5% sodium nitrate: dissolve 5 g $NaNO_3$ in 100 mL water.

EXPERIMENT 32

ISOLATION AND IDENTIFICATION OF DNA FROM YEAST

Special equipment

(12) Mortars
(12) Pestles
(6) Desk top clinical centrifuges (swinging bucket rotor)(optional).

Chemicals

(100 g) Baker's yeast, freshly purchased
(500 g) Acid washed sand
(1 L) Saline-CTAB isolation buffer: dissolve 20 g hexadecyltrimethylammonium bromide (CTAB, Sigma 45882), 2 mL 2-mercaptoethanol, 7.44 g ethylenediamine tetraacetate (EDTA, Sigma ED2SS), 8.77 g NaCl in 1 L Tris buffer. The Tris buffer is prepared by dissolving 12.1 g Tris in 700 mL water; adjust the pH to 8 by titrating with 4 M HCl. Bring the volume to 1 L.
(200 mL) 6 M $NaClO_4$ solution: dissolve 147 g $NaClO_4$ in water and bring the volume to 200 mL.
(100 mL) Citrate buffer: dissolve 0.88 g NaCl and 0.39 g sodium citrate in 100 mL water.
(1 L) Chloroform-isoamyl alcohol mixture: to 960 mL chloroform add 40 mL isoamyl alcohol. Mix throughly.
(2 L) Isopropyl alcohol
(50 mL) 1% glucose solution: dissolve 0.5 g D-glucose in 50 mL water.
(50 mL) 1% ribose solution: dissolve 0.5 g D-ribose in 50 mL water.
(50 mL) 1% deoxyribose solution: dissolve 0.5 g 2-deoxy-D-ribose in 50 mL water.
(200 mL) 95% ethanol.
(500 mL) Diphenylamine reagent. *This must be prepared shortly before lab use*. Dissolve 7.5 g reagent grade diphenylamine (Sigma D3409) in 50 mL glacial aceticacid. Add 7.5 mL concentrated sulfuric acid. Prior to use add 2.5 mL 1.6% acetaldehyde (made by dissolving 0.16 g acetaldehyde in 10 mL water).